Last Words

Last Words:
Large Language Models and the AI Apocalypse

Paul Kockelman

PRICKLY
PARADIGM
PRESS

Prickly Paradigm Press, LLC

5751 S. Woodlawn Ave.

Chicago, IL 60637

www.prickly-paradigm.com

ISBN: 9781734643558

Library of Congress Control Number: 2024945659

Printed in the United States of America on acid-free paper.

Contents

1.
The Horizon

The media has devoted article after article to large language models, such as OpenAI's ChatGPT, and the incredibly realistic, often eerie, and sometimes horrific conversations they can generate. A study claims that GPT-4, the "generative pretrained transformer" underlying the latest version of ChatGPT, can already pass its freshman year at Harvard. Another predicts that 340 million full-time jobs will be lost to artificial intelligence—based on the ability of generative AI to create content that is indistinguishable from human work. One article informs us that recent advances in artificial intelligence have made noninvasive mind-reading possible. Another instructs us how to turn our chatbot into a life coach. We learn that tech leaders have signed a letter calling for a "pause" in AI research and the creation of even larger language models because they fear the long-term negative effects. Only a few weeks later, the AI boom is generating optimism in the tech sector as "stocks soar."

And all that is already yesterday's news.

If language has long been an emblem of the human species, talking machines seem to be harbingers of some kind of technological singularity. Indeed, if the brilliance—or at least eloquence—of large language models is any indication, we seem to be poised at the threshold of general AI, a form of artificial intelligence that will not only meet, if not surpass, human intelligence but also maybe even replace humans altogether.

Messiah for some, Armageddon for others.

It is not just the recent acceleration of the verbal powers of such agents that is staggering, it is also their scale. They can contain trillions of parameters, require months of training, and have the entire corpus of the written word at their disposal. It is therefore tempting to call large language models, and ChatGPT in particular, hyperagents. This is not to say that such models are excessively agentive or so large and powerful as to be mysterious and unknowable. Nor is it just another way to fixate on their size. It is to stress that, whatever their actual capacities, they generate far more hype than other agents.

Perhaps only Jesus, Barbie, Obama, and T-Rex come close.

What follows is a spirited attempt to cram large language models into a relatively small text. I focus on the semiotic processes—or meaningful practices—that mediate the emergent relations among three kinds of actors: human agents, like you and me; machinic agents, such as ChatGPT; and corporate agents, be they states, tech companies, or institutions more generally. I pay particular attention to the coupling, and hence co-mediation, of the guiding principles of such agents. I thereby offer a critical genealogy of the highly contested relations among human values, machinic parameters, and corporate powers. My goal is not so much to see past our current social and technological horizon as to offer a theory of the reasons for and effects of the horizon itself.

Chapter 2, "Human Semiosis," offers an account of meaning that can span the distance between human interpretation and machinic calculation. Chapter 3, "Machine Semiosis," is a gentle introduction to the inner workings of large language models. Chapter 4, "Pretraining and Fine-Tuning," zooms in even closer, focusing on two key processes underlying the machinic mediation of meaning. Chapter 5, "Labor and Discipline," argues that language models are the objects and agents of disciplinary regimes and constitute a novel—and problematic—mode of production. Chapter 6, "Parrot Power," discusses the various kinds of generativity that underlie language models and shows their relation to linguistic competence, labor power, and corporate profits. Chapter 7, "Language Without Mind or World," discusses the limitations of large language models and users' limited awareness of those limits. Chapter 8, "Metasemiosis and Monsters," analyzes the mediatization of human-machine interaction and shows its relation to a range of enclosures, horizons, and scales. Chapter 9, "On Interpretation," takes up the distinction between humanistic and machinic interpretation and discusses strategies for interpreting texts that were generated by machines. Finally, chapter 10, "The Problem with Alignment," discusses the promises and pitfalls of aligning machinic parameters with human values, as well as the necessity of dealigning them with corporate interests.

Insofar as large language models, and machinic intelligence more generally, are a central target of speculative capital, they constitute a fast-moving topic. So rather than fetishize bleeding-edge developments, which usually only adds to the hype, I focus on bread-and-butter issues—especially in the lead up to ChatGPT. I write for a broad audience, and so for people willing to learn a little bit of math and work their way through a limited amount of formalism for the sake of a deeper understanding. I argue, implicitly, that critical theorists need to have a detailed understanding of the media they are analyzing—or else they are the equivalent of cavemen critiquing calculators. And I relate machine learning, language models, and natural language processing to meaning, and the great (post)humanist interpretive tradition, with a particular focus on ideas coming out of anthropology, critical theory, and pragmatism.

Given the stakes, as well as the hype, I will begin cautiously and work my way toward the horizon slowly.

2.
Human Semiosis

This chapter offers an account of meaning that can bridge the gap between humans and machines. It defines and exemplifies the key components of semiotic processes, shows the important role that values play in semiosis, and demonstrates how semiotic processes may embed and enchain. And, in preparation for the chapters that follow, it projects a simple functional notation onto human-specific semiotic processes so that they may easily be compared with those undertaken by large language models, and machinic agents more generally. In effect, I offer a noncanonical account of the grounds of interpretation, such that they may be extended past the limits of the human.

8

Components of Semiotic Processes

Figure 1 shows the key components of a semiotic process. Building on the ideas of Charles Sanders Peirce, the American logician and founding figure of pragmatism, a *sign* is whatever stands for something else. An *object* is whatever is stood for by a sign. An *interpretant* is whatever a sign creates insofar as it is taken to stand for an object. An *agent* is whatever can sense signs and instigate interpretants by way of relating to objects. Finally, *values* (which could also be called interpretative grounds or even guiding principles) are whatever an agent relies on to relate to signs, to relate signs and objects, and to relate objects and interpretants. Semiotic processes turn on motivation (what agents strive for) no less than meaning (what signs stand for).

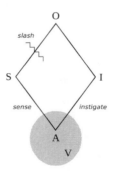

Figure 1. Semiotic Processes

Setting aside values for the moment, here are a few examples of semiotic processes. Someone (Agent) smells smoke (Sign), infers fire (Object), and calls for help (Interpretant). A telemarketer (A) hears the pitch of your voice (S), assumes you must be an adult male (O), and addresses you as "sir" (I). A student raises their hand (S), thereby indicating their desire to ask a question (O), and a teacher (A) calls on them (I). An interpreter (A) hears an utterance in French (S), which denotes a particular state of affairs and/or expresses a certain propositional content (O), which they then translate into German (I). Other examples include the exegesis of sacred texts, the diagnosis of illnesses, the undertaking of commands, the analysis of dreams, the explication of rituals, inferring a whole from a part, predicting subsequent events from preceding events, and far beyond.

In all of the foregoing examples, the interpretant makes sense in the context of the sign, given not just the interests (origins or identity) of the agent but also the features of the object (if only as imagined by the agent). As described in later chapters, the interpretant *aligns* with the sign insofar as it points toward the same object—however imprecisely.

While it is tempting to assume that objects are relatively objective and/or public (such as a tree that someone points to) and interpretants are relatively subjective and/or private (such as a thought or feeling), that is not necessarily the case. As these examples show, many objects are no more actual—and no less actual—than the desire one projects onto a person when they raise their hand; and many interpretants are as publicly available as signs, such as the German translation of the French sentence. And while many signs are communicative, insofar as they were intentionally expressed by an agent for the sake of securing an interpretant, the example of voice pitch highlights the fact that many sign-object relations—perhaps the majority—are nonintentional. The interpreting agent simply exploits (what seems to them to be) an existing correlation between a perceivable index and a putative identity.

As these examples also show, a key feature of many semiotic processes is the fact that the agent only learns about the object through the sign: a cause is known by its effect; an intention is inferred through an action; a desire is intimated by a gesture; an identity is revealed through an index; an illness is disclosed through a symptom; and so forth. Loosely speaking, there is something like a slash that separates the sign from the object: what can be directly sensed by the agent is on one side; what can only be indirectly known (by means of the sign) is on the other side. Notice the little wavy line in figure 1. As will be seen in later sections, the key slash in machine semiosis is not that which separates speech acts from mental states or states of affairs (and hence separates language from mind or language from world, as stereotypically understood) but that which separates earlier parts of a text from later parts, and/or the past from the future.

In effect, each and every semiotic process contains its own horizon.

Values as Guiding Principles

Agents rely on a wide array of resources to engage in semiotic processes. To get from the sign to the object, they might rely on the rules and vocabulary of a particular language (e.g., French). They might rely on a certain understanding of causality (e.g., fire leads to smoke). They might rely on certain social conventions (e.g., a raised hand indicates a desire to ask a question). And they might rely on certain projected patterns (e.g., men usually have deeper voices than women). Moreover, to get from the object to the interpretant, they might rely on certain ethical commitments and economic rationales (e.g., one should be brave, houses are valuable). They might rely on certain social norms or strategies (e.g., strangers should be addressed politely, especially if one is hoping to make a sale). They might rely on the rights and responsibilities associated with particular social statuses (e.g., teachers are obliged to answer the questions of students, time permitting). And so forth.

More generally, semiotic agents rely on their knowledge of the grammars and lexicons of particular languages. They rely on felicity conditions (in the tradition of John Austin): shared understandings regarding the appropriate and effective use of language in context. They rely on their theories, intuitions, analytics, paradigms, imaginaries, hermeneutics, causal logics, epistemes, and worldviews. They rely on their taxonomies, partonomies, ontologies, schema, scripts, frames, stereotypes, prejudices, and biases. They rely on shared norms, rules, laws, conventions, protocols, and traditions. They rely on the affordances of various materials and/or the technological constraints of various media. They rely on their understandings of minds, signs, media, technology, language, nature, self, and society. They rely on morals, ethics, ideals, and evaluative standards. And, of course, they rely on context, cotext (meaning co-occurring text), and culture. Indeed, culture itself might be understood as relatively shared values, constituting something like the semiotic commons of a particular collectivity of agents.

Such interpretive resources, or *values*, are fundamental to semiotic processes. Understood as agent-specific sensibilities and assumptions, they function as guiding principles that allow agents to interrelate objects, signs, and interpretants. As such, they constitute the grounds of attention, affect, action, and inference. In particular, values help determine:

· what an agent notices (such that it might constitute a sign in the first place);
· what an agent infers or otherwise comes to know (given the sign so noticed);
· how an agent acts, thinks, or feels (given the object so known).

Values may be encoded in texts; embodied in habits; enminded in beliefs and desires; embrained in neural networks; embedded in infrastructure, artifacts, and environments; and even engenomed in particular species. And many disciplines have long analyzed the genealogy of such values: the history of their creation, transformation, stabilization, and spread. In what follows, I will usually focus on values that are group-specific and historically changing and not be too concerned with their discipline-specific elaboration.

Although values often remain in the background of semiotic processes (which tend to be more noticeable figures, insofar as such processes involve relatively public actions and utterances), they can easily become figured. In particular, the objects of semiosis are often the values that guide semiotic processes: agents can topicalize, characterize, and reason about their values. In this way, agents can communicate and critique their own and others' values.

Moreover, values are not only a condition of possibility for and the objects of semiotic processes, they are often the consequences as well as the ends of semiotic processes. Indeed, many interpretants are precisely changes in habits and beliefs, or values more generally, and hence changes in an agent's propensity to interpretant future signs in particular ways.

In short, semiotic values are dynamic variables: at once the objects and interpretants, as well as the roots and fruits, of semiotic processes. As will be seen in the chapters that follow, they also play a decisive role in the machinic mediation of meaning.

Embedding and Enchaining

Apropos of the last set of points and looking forward to later arguments, I foreground two frequently occurring modes of semiotic mediation.

Figure 2 shows how semiotic processes may *enchain*: the interpretant in a prior process may constitute the sign in a subsequent process. For example, when a teacher calls on a student (as an interpretant of their having raised their hand), that itself also constitutes a sign (indicating that the student may now ask their question).

Such enchained semiotic processes constitute the backbone of everyday interaction and play a central role in mediating social relations. In particular, the relation between the sign and the interpretant (as two entities or events, with coupled

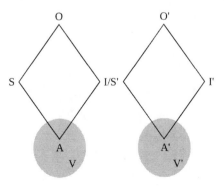

Figure 2. Enchaining

relevance) mediates the relation between the signer and the interpreter (as two agents, with complementary and often emergent identities).

Figure 3 shows how semiotic processes may *embed*: the object of a process may be constituted by any component of a process, any relation between such components, or any enchaining of semiotic processes more generally. To build on the previous example, another student in the class may later use reported speech, or even a stick figure cartoon with word balloons, to capture the interaction between the teacher and the student.

The values underlying semiotic processes are often the objects of semiotic processes: we can describe not just how the teacher responded to the

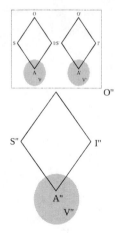

Figure 3. Embedding

student but also how they should have or could have responded. What we represent, and otherwise signify and interpret, is very often that which mediates our modes of signification and interpretation.

The Horror

As a kind of shorthand going forth, it will sometimes prove useful to denote human semiotic processes in a quasi-functional notation:

$$I = A_V(S) \tag{1}$$

Such a notation likens a human semiotic agent to a mathematical function, A. The input to this function is some sign, S; the output of this function is some interpretant, I; and the parameters of this function are the values of the agent, V. Compare a simple linear function like $y = f_\theta(x) = mx+b$, where x is the input, y is the output, and $\theta = \{m,b\}$ is a set of adjustable parameters that determine the slope and y-intercept of the line in question.

The point here is not to determine such a function, and certainly not to suggest that human semiotic agents are constituted by such a function (even if they may be modeled as one) but rather to condense all the dirty details of semiotic processes as they unfold in the wild, so to speak, in a compact notation for the sake of later comparison.

Humanists will, no doubt, be horrified. But I thought I might, in light of what comes next, meet the machines halfway.

3.
Machine Semiosis

This chapter compares the key components of machine semiosis with those of human semiosis and shows how the parameters of machines are coupled to, and thereby made to align with, the values of people. By offering a gentle introduction to the mathematics underlying such models, I aim to dispel some of the mystery—and magical thinking—that otherwise surrounds them.

Large Language Models

At a certain level of abstraction, a large language model may be understood as a parameter-dependent function that accepts a sequence of words as its input and returns a sequence of words as its output. Assuming the function was well chosen and its parameters have been adequately set, the inputted sequence, known as the *prompt*, specifies a task that the user wants fulfilled, and the outputted sequence, known as the *response*, fulfills that task.

For example, if the prompt is "alphabetize the following words: bat, dog, cat, zebra, armadillo," the response could be "armadillo, bat, cat, dog, zebra." If the prompt is "translate the following sentence into English: me llavo las manos," the response could be "I wash my hands." If the prompt is "what is Napoleon most famous for?," the response could be "conquering much of Europe." And if the prompt is "write a short story, in the style of Chekhov, involving three clowns and a cabbage," the response would be just such a story. Other actions a large language model may be asked to undertake include offering lifestyle tips, writing algorithms, brainstorming, extracting evidence from texts, and the like.

Recursively, and more generally, if the human user inputs a discursive move, the language model can output a felicitous response to that move, which can itself constitute a discursive move calling for its own response (recall the example of semiotic enchaining), such that a language model can engage in human-like conversations. This is what puts the *Chat* in ChatGPT.

More carefully, the prompt is typically a sequence of words that describes, and perhaps demonstrates, certain satisfaction conditions, and the response—at least when all goes well—is a sequence of words that satisfies those conditions. Phrased another way, if prompts are descriptions of actions that the user wants the model to undertake, responses are the results of the actions so undertaken.

In short, at this level of abstraction large language models are very complicated semiotic agents, with prompts as their signs, responses as their interpretants, satisfaction conditions as their objects, and parameters rather than values as their guiding principles. See figure 4 (and recall figure 1).

But unlike the example of human semiotic agents discussed in the preceding chapter, such

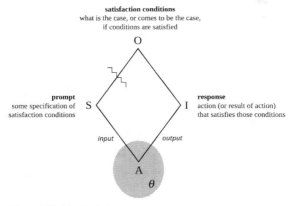

Figure 4. Machine Semiosis

models really are mathematical functions. And so, rather than saying that a machinic agent senses signs and instigates interpretants, it is better to say that such an agent accepts signs as inputs and returns interpretants as outputs. Its output depends not just on its input but also on the mathematical details of the function in question, as well as the particular numerical values of all its parameters.

In a certain sense, then, a large language model is simply a mathematical function that behaves in a human fashion. How was it disciplined to do so?

Pretraining and Fine-Tuning

Such a movement from prompt to response, which involves the calculation of a function's output when given an input (and already established parameter values), is known as *forward propagation*. It may be summarized as follows:

$$I = A_{\theta}(S) \tag{2}$$

But before a language model can respond to prompts in a way that satisfies the desires of its users, its parameters (θ) must be set. This means determining good values for potentially trillions of variables by training the model to undertake certain carefully chosen tasks. The model is repeatedly given various inputs, and the values of the parameters are slowly adjusted, through an algorithmic process known as *backpropagation*, until the outputs relate to those inputs in a way that is deemed adequate for the tasks in question.

While there are many such tasks, two stand out in terms of their overall importance for a large language model like ChatGPT. In a process known as *pretraining*, the model is given sequences of words from a huge corpus of human-authored texts and asked to predict the next word in the sequence. Such training gives language models their distinctive ability to produce next words, conditioned on prior words, and thereby to *generate* sequences of words, or texts, that seem formally cohesive and functionally coherent. This is what puts both the *P* and the *G* in ChatGPT.

In short, pretraining a machinic agent to mirror the actual makes it good at generating the plausible.

Language models are surprisingly capable with only pretraining (given enough parameters, training data, and computational effort), but the word sequences outputted only really relate to the word sequences inputted as textual continuations. For that is all the models were trained to produce. To make the outputs consistently relate to the inputs as responses to prompts, and hence as the semiotic satisfaction of the stated conditions (as described above), *fine-tuning* must take place.

While there are many varieties of fine-tuning, the most important kind is arguably reinforcement learning with human feedback. It involves several steps. First, human judgments are used to rank possible responses to various prompts in terms of their relative preferability (given some standard of values). For example, which of two responses is considered more helpful, truthful, and harmless? Second, those rankings are used to train a second language model (known as a "reward model") to output numerical scores consistent with those rankings when given prompt-response pairs as inputs. That is, a reward model is trained to numerically mirror human preferences regarding the relative helpfulness, truthfulness, and harmlessness of responses. Third, the outputs of this second language model are used as a reward mechanism, or feedback signal, to further train the original model (that was initially pretrained to engage in next-word prediction), such that the responses it produces better satisfy the prompts of its users. Just as a machine learning algorithm may be trained to play a video game (by acting on its environment in a way that maximizes its score), a language model is thereby trained to play language games (by responding to users' prompts in a way that maximizes its reward).

In short, for machinic responses to align with human prompts during forward propagation, machinic parameters must be made to align with human values during backpropagation (through pretraining or fine-tuning). See figure 5.

Slashing Words from Worlds

Recall, from chapter 2, that slashes separate signs from objects, and hence something like the perceived from the intuited, what is present from what is absent, or what is given from what is inferred. Language models involve many such slashes. At a

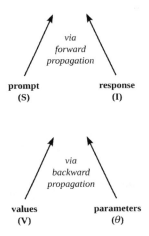

Figure 5. Two Modes of Alignment

relatively high level of abstraction, as was shown in figure 1, the slash between sign and object separates prompts from satisfaction conditions, or speech acts from communicative intentions. At a lower level of abstraction, as seen in the context of next-word prediction during pretraining, the slash separating sign and object separates earlier words from later words, and hence something like the past from the future.

In particular, next-word prediction, itself a key part of response formulation, is also a semiotic process. See figure 6. The sign is a sequence of words from a human-authored text. The object is the actual next word in the sequence (as it occurs in the text). And the interpretant is a probability

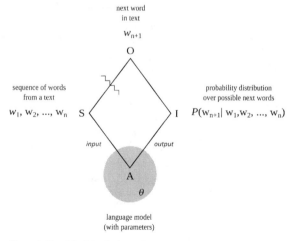

Figure 6. Next-Word Prediction as Semiotic Process

distribution over possible next words (conditioned on the preceding words).

At this level of abstraction, the slash separating sign and object, and hence that which separates what is given from what must be inferred, is not the kind of slash that separates representations from the world, appearance from identity, performance from competence, action from intention, or words from things (as stereotypically understood). It is, rather, the kind of slash that separates the future from the past, or earlier parts of texts and scores from later parts. And hence it turns on interpretive grounds that are similar to those that govern musical expectations and poetic prefigurings, such as repetition, parallelism, meter, echos, and refrains.

Positively framed, predicting what comes next, given what has come before, is the fundamental capacity of such machinic agents. Negatively framed, large language models traffic in word-word relations, not word-world relations.

As will be seen in later chapters, this positioning of the slash both animates and haunts such agents.

Parasitic Intentionality

By referring to such language models as semiotic agents, I am not trying to project sentience or sapience, or any other aspect of human subjectivity, onto them. Rather, I am simply foregrounding the fact that such models are capable of engaging in what seem to be complicated acts of semiosis and embody (in their functional architecture and numerical parameters) a mode of intentionality that is derivative of their makers.

In particular, large language models are mathematical functions that serve instrumental functions derived from the purposes of the human agents who created and trained the models in question (such that a model's interpretants of particular signs, qua outputs, come to more and more closely resemble human interpretants of the same signs).

Indeed, the intentionality (or object-directedness and ends-directedness) of the trained model is derivative not just of the intentionality of the humans who trained it (insofar as they want to make a machine that serves a certain instrumental function by creating a machine that calculates a certain mathematical function) but also of the intentionality of the humans who produced the texts and instructions it was trained on (such as corpus data, prompt-response pairs, preferability judgments, and alignment criteria).

In this sense, the signs and interpretants of language models relate to something outside of themselves, and thereby possess intentionality, by virtue of being parasitic on a more originary mode of human intentionality (which is itself derivative of processes like natural selection, not to mention education, enculturation, and indoctrination).

That said, I will later investigate why it is so easy, and perhaps alluring, to project complex capacities like consciousness and choice onto such agents, so that they might come to be not just personified but also fetishized, and perhaps even deified, by unsuspecting human agents.

4.
Pretraining and Fine-Tuning

This chapter examines pretraining (for next-word prediction) and fine-tuning (for aligning with users' intentions) in greater detail. It thereby offers a closer look at the coupling of human values and machinic parameters. More colorfully, it examines the machinic disciplinary regime that brings a novel kind of discursive agent into being.

Next-Word Prediction

Figure 7 summarizes the key operations a language model undertakes during forward propagation, when it generates plausible stretches of text by means of next-word prediction.

The input to the model (some swatch of text) is first parsed into a sequence of "words," or tokens more generally. Such tokens include words proper and also parts of words, punctuation marks, whitespace characters, end-of-sequence markers, and the like.

Each of those words is then translated into a distinct vector, known as a *decontextualized word embedding*, which represents the meaning of the word as a long list of numbers and hence as a position in a relatively abstract, high-dimensional space.

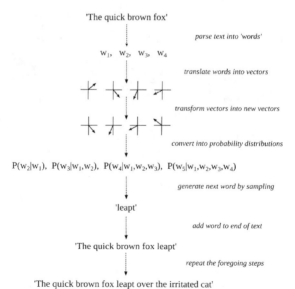

Figure 7. Word Prediction and Text Generation

Two words are similar in meaning, in the sense that they can play the same role in a sequence of words (for the sake of next-word prediction), if their associated embeddings constitute "nearby" positions in that multidimensional space. For this to happen, the model has something like a dictionary that maps words to word embeddings and thereby translates words into lists of numbers. See figure 8. This is where the semantic meaning of words, as it were, without reference to their syntactic position in a sequence of words, enters the process.

A celebrated function, known as a *transformer*, is then repeatedly applied to this sequence of vectors. It is what puts the *T* in ChatGPT. The inner workings of transformers are quite complicated and mainly involve a lot of matrix multiplication (where a matrix may be understood as a two-dimensional array of numbers). Multiplying a vector by a matrix results in another vector that relates to the original as some kind of transformation (like a rotation,

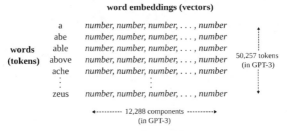

Figure 8. Dictionary, as a Mapping from Words to Vectors

stretch, or distortion). Crucial to such transformations, every vector is made to attend to and interact with the vectors that precede it in the sequence. The final result is a new sequence of vectors, known as *contextualized word embeddings*, each of which should now represent the meaning of the *next* word in the sequence (conditioned on, and hence mediated by, the words that came before it). This is where grammatical structure, as well as more long-distance textual relations, enter the process.

Each of these transformed vectors is then compared with all the word embeddings in the dictionary (themselves just vectors). The closer a transformed vector is to any such word embedding (that is, the more it points to the same position in that multidimensional space), the greater the probability that is assigned to the word associated with the embedding as the next word in the sequence. Each transformed vector is thereby converted into a probability distribution (over all possible words in the dictionary of the model). See figure 9. The first probability distribution in the series, $P(w_2|w_1)$, predicts the second word conditioned on the first word. The second probability distribution in the series, $P(w_3|w_1,w_2)$, predicts the third word conditioned on the first and second words. And so forth.

The last probability distribution in this series, which represents the probability of a word that was *not* in the original textual input, conditioned on all the words that were in the input, is then used to randomly generate a plausible next word in the sequence. That is, the greater the probability assigned to a word in the final distribution, the more likely it will be chosen—or rather "rolled"—as the next word in the sequence.

The randomly chosen word is then added to the original sequence, and the entire process is repeated, again and again, until a special "word" (such as an end-of-sequence token) is generated. The newly generated sequence of words (without the original textual input) is then outputted, as the textual continuation of the input. For example, if "the quick brown fox" goes into the machinic agent, "leaped over the irritated cat" could come out.

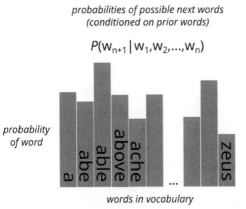

Figure 9. Probability Distribution

The terms *contextualized* and *noncontextualized*, as applied to word embeddings, are misnomers. In no sense is context being taken into account, if context is understood as the conditions in which such texts (or the sequences of words within them) were said, written, read, thought, or otherwise signified and interpreted. Rather, what is taken into account is *cotext*: the way that the meaning of each word is mediated by the other words that came before it in some swatch of text. Recall figure 6. I use the expression "context" rather than "cotext," because that is the norm in the natural language processing community. But, as should be clear, that is radically optimistic, if not downright fanciful.

As amazing as large language models are at taking cotext into account, they are, as of yet, minimally able to take into account context—and hence the immediate environment, speech event, conversational background, or world more generally.

Parameters

The emphasis so far has been on how a pretrained language model generates formally cohesive and functionally coherent text using already determined parameter values. The focus has been on forward propagation (without fine-tuning). Several interrelated questions may now be answered: Where exactly are the parameters in the model, what is their function, and how were they determined via backpropagation?

Simply stated, the parameters are all the numbers in the word embeddings and matrices mentioned above. Such numbers represent either the meanings of words (as positions in high-dimensional spaces) or the structure of mathematical transformations that may be applied to such vectors (such that the transformed vectors come to represent next words conditioned on prior words).

At the onset of training, all those numbers are randomly initialized. With pretraining, the model is fed sequences of words from human-authored texts. Rather than generating new words using only the final probability distribution (as just described), the model compares each of the probability distributions in the series with the actual word that comes next in the human-authored text at that point in the sequence. In other words, $P(w_2|w_1)$ is compared with

w_2, $P(w_3|w_1,w_2)$ is compared with w_3, and so forth. And the parameters of the model are slowly adjusted to make such probability distributions better and better at predicting the actual words that come next in the sequence. Technically speaking, the goal is to minimize a function known as *cross-entropy loss*, which is more or less equivalent to maximizing the model's predictive accuracy.

Given that each of its word embeddings consisted of 12,288 numbers and there were 50,257 words, or tokens more generally, in its vocabulary, a large language model like GPT-3 had 50,257 x 12,288 parameters devoted to word embeddings alone, and thus over half-a-billion floating-point numbers devoted to vocabulary items (or lexical semantics, so to speak). Adding all its other parameters, such as all the numbers in those transformative matrices, brings its parameter count up to about 175 billion. And newer models are much larger.

This should give readers a sense of what the modifier *large* really means when used in an expression like "large language model." In contrast to the hype surrounding generalized artificial intelligence, the application of this adjective to language models to capture the scale of their parameter space is relatively *hypo*bolic.

The next section takes up fine-tuning, whereby the parameters of the model are further adjusted until the outputted text relates to the inputted text not just as a textual continuation but also as a response to a prompt—and hence as a fully fledged interpretant of a sign.

Alignment Through Fine-Tuning

Suppose that a language model, A_θ, has been sufficiently pretrained as just described. The parameters of the model, θ, have thus been set in such a way that it can recursively engage in next-word prediction, and thereby "continue" any text it is given. So that such a machinic agent can consistently respond to prompts in ways that seem to satisfy both the intentions of its users and the interests of its creators, fine-tuning must be done. There are many varieties of fine-tuning, each designed to give language models distinctive abilities (above and beyond next-token prediction per se). This section will focus on reinforcement learning with human feedback, given its importance to the abilities of large language models like ChatGPT and its relevance to the alignment problem more generally.

To begin this process, a set of prompts, along with possible responses to them, is collected or created. As discussed in chapter 3, the prompts are typically descriptions of tasks that users would like a language model to undertake, and the responses are typically the tasks so undertaken. For example, the set might include a wide variety of questions and commands as prompts and for each, several responses of varying quality, such as answers to those questions and undertakings of those commands.

The prompts themselves may be based on input from past users of language models, and apps more generally, regarding what kinds of questions and commands are frequently used or considered important. The responses may also be written by human agents or harvested from the internet but are usually generated by the language model itself. In particular, each of the prompts is given to the pre-trained model several times, and its various outputs (as possible responses to the same prompt) are collected.

This first stage of the fine-tuning process requires human labor to find or create a set of prompts, as well as machinic labor, itself grounded in human labor, to produce responses to them. The next stage requires human labor to rank possible responses to the same prompt in terms of their relative preferability. This is a key place where "human feedback" explicitly enters the training process.

Suppose, for example, that prompt P is a question and responses R_1 and R_2 are possible answers to that question (as generated by the pretrained language model). Human judgment (in the form of paid evaluators, usually hired on short-term contracts) is needed to decide whether R_1 is more preferable than R_2 ($R_1 > R_2$), R_2 is more preferable than R_1 ($R_2 > R_1$), or both responses are equally preferable (meaning the people are undecided). As should be clear, such comparisons are very similar to the types of preference relations analyzed by economists to model a person's values in terms of a utility function; but now such comparisons are applied to the interpretants of signs rather than to commodity bundles per se.

Of particular importance during this stage is the establishment of a set of *alignment criteria* that specify what counts as a good response to a prompt in the first place, such that any two responses to the same prompt can be ranked in terms of their relative preferability. There is a lot of literature, as well as debates, around these issues. The following criteria often come up:

· Responses should satisfy the intention of the person who provided the prompt. For example, if a user asks a question, the response should answer that question. If a user commands an action, the response should undertake that action. In effect, a machinic agent capable of satisfying communicative intentions must first be able to recognize such intentions and hence be able to identify the illocutionary force and propositional content of the prompt. Is the prompt a question or a command, or some other kind of speech act entirely? And what, in particular, is being asked, commanded, or otherwise requested, however elliptically? This criterion is sometimes described as being "helpful."

· As part and parcel of being helpful, responses to prompts should also be truthful. In other words, responses should adhere to "the facts" insofar as such facts are relevant and established. Sometimes this criterion is couched as being "honest," but that way of wording it presumes that language models have mental states that may or may not be aligned with their speech acts: e.g., whether or not they really believe what they say. To be sure, given the current abilities of large language models, some readers might expect that sincerity criteria will need to be added soon enough: believe what you say; intend what you promise; regret what you apologize for; and so forth. In any case, the important issue is that responses conform to the facts (as understood by those who are evaluating them).

· Responses should not just be helpful and truthful, they should also avoid bias, not contain sexual or violent content, not use toxic language, not denigrate protected classes of people, not provide information that could prove harmful (e.g., the instructions for building a bomb), not pass themselves off as more capable than they are (e.g., they should remind the reader they are simply a large language model, not a sentient being), and so forth. Such a list of requirements could be extended indefinitely, and what should or should not be on it is subject to intense debate. This criterion is sometime phrased as being "harmless."

· Finally, many criteria could be added to this list that have less to do with satisfying the intentions of users of language models and more to do with satisfying the interests of the makers of such models, or the owners and creators of any downstream apps that may incorporate the models. Within this set are not just all the foregoing criteria (for it is often in the interests of such corporate agents to satisfy the intentions of their customers) but also additional criteria (that may be understated in publicly available technical reports). For example: do not break any laws, make sure responses are likely to bring users back, stoke desire for our product, paint a rosy picture of a certain worldview.

Linguists and philosophers will, no doubt, hear echos of Paul Grice's famous conversational maxims (make your contribution to a conversation be informative, truthful, relevant, and clear), which to a certain degree simply mirror the prescriptive urgings of parents and teachers. Others will hear echos of John Austin's felicity conditions (contributions to discourse should conform to shared understandings of what counts as an appropriate and effective utterance in the current context). And still others will hear echos of the Ten Commandments or the Golden Rule. But such maxims and conditions were just philosophers' intuitions regarding the workings of language, or the purported desires of deities regarding the behavior of their followers. In the case of large language models, in contrast, people are paid to rank responses according to the above criteria so that the discursive behavior of large language models can be made to conform to such criteria (at least to a certain degree). In this way, the behavior of machines can be made to better align with the values of people and the interests of corporations.

In short, the distance from say unto others as you would have them say unto you, to maximize shareholder value by any means possible, is but a step.

Figure 10 highlights the recursive nature of these evaluative judgments. Just as human agents can rank responses to prompts as a function of the degree to which they satisfy certain alignment criteria (e.g., are they helpful, truthful, and harmless), corporate agents can rank (hire and fire) human agents as a function of the degree to which they follow instructions (regarding how to rank responses to prompts). In other words, just as a company wants the responses of its language model to be aligned with the prompts of its users, it wants the judgments of its hired help to be aligned with its instructions.

Now comes the next important step in the fine-tuning process. If a language model takes a prompt as its input and returns a response as its output, a reward model, A^{rm}_{ϕ}, takes a prompt-response pair as its input and returns a numerical value as its output. All the alignment criteria discussed above, which reflect the values of certain "people" (for better or for worse), are thereby condensed into a single number. As such, an evaluative standard is rendered quantitative and monodimensional.

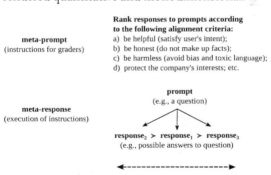

Rank responses to prompts according to the following alignment criteria:

meta-prompt
(instructions for graders)

a) be helpful (satisfy user's intent);
b) be honest (do not make up facts);
c) be harmless (avoid bias and toxic language);
d) protect the company's interests; etc.

meta-response
(execution of instructions)

prompt
(e.g., a question)

response₂ > response₁ > response₃
(e.g., possible answers to question)

responses ranked by their degree of alignment with the prompt
(according to the alignment criteria stated in the meta-prompt)

Figure 10. Meta-Alignment

Reward models are typically pretrained language models, as large and complex as the language model being fine-tuned, that have been tweaked to return a single number rather than a sequence of words as their output. As may be seen from the subscript Φ, they have their own set of parameters to train. Crucially, such reward models are themselves fine-tuned so that the numbers they return scale with, and thereby mirror, the preference relations of people. In particular, if the evaluators ranked response one (R_1) as more preferable than response two (R_2), given some prompt (P), then the model is trained to output a higher number for response one than for response two in the context of that prompt:

$$\text{if } R_1 > R_2 \text{ then } A^{rm}{}_\Phi(P,R_1) > A^{rm}{}_\Phi(P,R_2) \qquad (3)$$

In other words, after training, the numerical outputs of this machinic agent should match the preferences of the human agents, whose judgments should match the alignment criteria of the corporate agents that are training the language model.

In short, if a language model is trained to speak in a human-like fashion, a reward model is trained to evaluate the speech of language models in a human-like fashion. It becomes a metalanguage model, and hence a metalinguistic agent, able to recognize and evaluate not just the syntax and semantics but also the pragmatics of language models. Hence it is a kind of meta-evaluative and meta-interpretive machinic agent that can provide feedback on the outputs of a language model. Phrased in terms of social relations, we now have a machinic teacher who can grade the responses of a machinic student, to a wide variety of prompts, and thereby flatten context-specific satisfaction conditions into a single numerical score.

Once a reward model, A^{rm}_{ϕ}, has been trained using human feedback, the original language model, A_{ϕ}, can finally be fine-tuned using reinforcement learning. The key algorithm underlying this last step (known as PPO, for proximal policy optimization) is quite complicated; the overall logic may be summarized as follows. Give the language model a prompt and collect its response. Give the prompt-response pair to the reward model and collect its score (a number that scales with the preferability of the prompt-response relation). Use that number as a reward, or feedback signal, for the language model, an indication of how well the model is doing with its current parameters. Adjust the parameters of the language model via backpropagation so that its response to the prompt would have received a higher score. Finally, repeat this process over and over until the language model consistently produces high-scoring responses to prompts, responses that human evaluators would find preferable given their alignment criteria.

Just to be clear, the language model still engages in next-token prediction. However, its parameters are tweaked not to make it predict next words more and more accurately but so that its final outputs (as responses to prompts) achieve higher and higher scores from the reward model. This, of course, is why the process is called reinforcement learning: it algorithmically embodies the principle that behavior that was rewarded in the past is more likely to be repeated in the future.

This procedure is very similar to the way that machine learning algorithms are trained to play games like chess. However, rather than being trained to engage in certain actions (such as moving a knight) as a function of its current environment (understood as the current positions of all the other pieces), the machinic agent is trained to produce the next word as a function of the preceding words. In effect, it resides in and acts on a textual environment. And rather than being trained to win a game like chess or get a high score in a game like Pac-Man, it is trained to get a high score from the reward model for its response, and thereby do well when playing a particular kind of language game.

A language game—or, rather, a mode of language gamification—with ethical grounds, economic rewards, and existential risks.

5.
Labor and Discipline

Having described the inner workings and training regimes of large language models, it is useful to frame their historical emergence and future potential in several related ways: they are the product of particularly intense forms of labor and the effect of particularly severe modes of discipline; and as a function of all this, they not only are capable of speaking (so to speak) and engaging in machine semiosis per se, they also have the potential to perform labor and discipline others. Indeed, they will engage in both kinds of activities for the sake of humans and in place of humans and make humans the subjects, if not the targets, of such activities. Let me unpack these points, focusing first on the relation between labor and language models.

Language Models and Laboring Subjects

Training a language model through backpropagation, for the sake of next-word prediction or alignment more generally, may be understood as a mode of labor or a type of work in four overlapping senses. First, backpropagation creates both a use value and an exchange value, and hence is a process that is both concretely and abstractly productive. More specifically, the large language model itself, with its generative and predictive capacities, is a utility that seems to satisfy human needs and desires. And such a machinic agent, once packaged and otherwise made portable, is a commodity that can be bought and sold.

Building on this last point, backpropagation gives form to substance for the sake of function, and thereby turns relatively raw materials into an (almost) finished product. Here the substance is constituted by the model itself, prior to training, and so with all its parameter values only randomly initialized. The form consists of better parameter values, as achieved through backpropagation. And the model acquires novel functions insofar as its capacity to generate, predict, and respond is improved, and insofar as the price it commands as a good or service is increased.

Like any other productive activity, training a language model consumes a huge number of resources

—not just the time and labor of those involved, but also water (for cooling servers) and energy (for carrying out the calculations that determine all those parameter values). It also produces all sorts of waste products (and hence 'bads' as opposed to goods), from heat to CO_2 emissions. And while the cost of training is astronomical, it is dwarfed by the cost of actually running the trained models (to respond to queries, and thereby interpret signs). Indeed, the energy requirements of such models are so high that many experts believe they will be the key bottleneck on future progress, as well as a key factor in upcoming geopolitical struggles.

Finally, training a model through backpropagation organizes complexity (the state space of all possible parameter values) for the sake of predictability. This is similar to the way compressing a container of gas organizes complexity (the space of all possible positions of molecules) and thereby creates predictability (the molecules become localized in a smaller volume, such that the disorder—or entropy—of the gas is decreased). Doing work on the gas creates an agent that is itself capable of doing work. For the compressed gas, like a stretched spring, is now primed to lower its pressure by increasing its volume, and thereby do work on its environment.

In other words, training a large language model involves something like work (the movement of a force through a distance, the expenditure of energy, the emission of waste products) and hence a struggle against disorder (in particular, the effort it takes to reduce cross-entropy loss and thereby improve predictive accuracy), all in the service of creating an agent capable of doing work: in particular, *the work of interpretation*.

To be sure, and looking slightly ahead, language models are not only made through various modes of labor (and so constitute a product) but also used to make other products (and so constitute an instrument), and they are even that which makes (and so constitute a laborer, if not labor power per se). Not only do they carry value (by being a commodity that can be bought and sold), they can arguably create value (through their labors), and they will soon be able to realize value (by buying and selling other commodities on their own or in our stead). They and their brethren are capable of analyzing events in order to identify patterns that can be profited from—if only patterns of speaking and, through those, of culture and of desire. They can make many other modes of labor obsolete, in particular, all the activities currently undertaken by all the workers they will replace. And they can be used to siphon value out of a system (playing the role of a middleman or parasite). Finally, their creation arguably relies on stolen property: all those human-authored texts they were trained on without acknowledging the original authors or giving anything back in return.

In short, large language models play a number of decisive—and arguably devastating—roles when seen through the lens of critical political economy.

Machines as Disciplined Subjects

Insofar as it creates a machinic agent, capable not just of generation and prediction but also of signification and interpretation more generally, training a large language model should also be understood as a mode of discipline, control, or governance. To see how, consider the following points.

Given a sign, a language model produces an interpretant (which itself constitutes a sign for further interpretants, and hence a prompt for future responses). As was shown earlier, such models thereby behave in a way that can be brought into alignment not just with particular human practices but also with the values that constitute the guiding principles underlying those practices. Phrased another way, training enlists one agent (the back-propagation algorithm) to channel the behavior of another agent (the language model itself) into more appropriate, desirable, and exploitable forms.

Concomitantly, such modes of regimentation bring unruly tokens into alignment with normative, or otherwise legislated, types. In other words, the generative capacity of a language model constitutes a kind of linguistic competence or discursive power. And such a competence is subject to a range of controlling processes such that the performance of that competence or the exercise of that power becomes more and more grammatical, felicitous, rule-abiding, normative, ethical, profitable, and malleable. (At least when judged in light of prevailing norms, ethical standards, or models of desirable comportment that hold within certain collectivities.) Indeed, besides the language model, as an agent, being made predictable (itself a key criterion in many accounts of subject formation), the model is made capable of making predictions in more and more desirable ways, such that its actions can be not just directed but also capitalized on and constrained.

Finally, all these forms of governance have as their effect, or emergent product, a kind of quasi-subject: that which thinks and speaks; that which can represent and be represented; that which can be the subject and object of paradigms and epistemes, not to mention the player and umpire of language games; and that which, soon enough, can be not just the author and instigator but also the principal and benefactor of discursive actions.

Or so it seems. For, as will be seen in later sections, despite their incredible capacity to "speak," language models are often as dumb as can be.

Machines as Disciplining Agents

The focus so far has been on a range of processes, involving both work and discipline, that bring machinic parameters into alignment with human values: θ => V. The direction of mediation can run the other way, such that human values are brought into alignment with machinic parameters: V => θ. At the risk of adding to the hype, this section describes some of the ways such realignment and dealignment may happen.

Language models have long been used to suggest next words when we write and text. And algorithms, in the service of language processing, have also been used to check our spelling, edit our grammar, organize our essays, suggest synonyms, point out clichés, speed up our search queries, and the like.

As may be seen with platforms like Khanmigo and Duolingo, large language models will be incorporated into a variety of applications to educate children and adults across the globe: not just to speak their own languages in more standardized ways and to learn other languages, but to learn just about any other subject that can be taught and tested. The movement from education to indoctrination, like the movement from knowledge to ideology, or nudging to coercing, can be subtle and shifting.

Large language models will function not just as teachers and editors but also as analysts, advisers, brokers, gurus, therapists, strategists, oracles, sidekicks, detectives, interrogators, ethnographers, and superegos. They will guide us through important decisions, help us interpret our behavior in light of our upbringing, figure out what we value or how we reason, and even berate us for having acted, felt, or texted as we did.

Even more pessimistically, they will be used more and more to oversee and discipline humans: tracking what we have said and done, predicting what we will do and say next, telling us who is right, how to vote, what to buy, and even whom to save, ignore, or kill.

They will, in particular, be used to generate texts (new stories, propaganda, memes, advertisements, philosophies, cosmologies, myths, distractions, screenplays, and conspiracies). And such texts not only will change our values in relatively indirect ways but may also ensure that we come together less often, and in less democratic ways, to agentively determine our own values relatively directly—such as lowering the probability that people participate in forums in which they disclose and debate shared principles that could guide their collective actions.

Indeed, not only will language models be a source of signals (however uninformative, dishonest, or false), they will also be a source of noise, or a parasite more generally. They will intercept our messages (by diverting them to unintended agents or deciphering them along the way). They will interfere with messages (by distorting their contents, reducing their informativeness, and/or degrading their truth value). And of course, they will come to create so many new messages, or "texts" (including scientific reports, opinions, and newspaper articles) that nobody will know who wrote what, which texts are worth reading, or what should be believed.

All the foregoing processes will come to affect deeper and deeper aspects of human subjectivity: the beliefs people have, the things they hold dear; their affect and intentions, dreams and habits, subconscience and unconscience; how they represent the world, and who they want as their representatives. And, following the arguments of chapter 2, insofar as people's values are transformed in these ways, so too are their semiotic processes to the extent they are guided by those values: what people notice or otherwise attend to; what people infer or intuit from what they notice; and how people act, and are otherwise affected, by their inferences and intuitions.

In short, just as one can offer a political economy of machinic agents, one can offer a genealogy of their parameters. And just as training a large language model brings into being a novel kind of subject, with distinctive modes of agency, such models—by training, or at least entraining, human beings—will decisively transform older forms of subjectivity and may come to lessen, if not altogether diminish, foundational modes of human agency.

To be sure, most of the processes just mentioned have long been underway, as evinced in older forms of media—including language itself. When mediated by large language models they will arguably be scaled up, commodified, and weaponized in unprecedented ways.

6.
Parrot Power

The preceding chapter showed how machinic parameters are mediated by human values and, in turn, how human values are mediated by machinic parameters, sketching the political economy and genealogy of machinic subjectivity. This chapter examines the characteristic power, or competence, of large language models: their generative capacity. It theorizes various kinds of generativity and shows their interrelation. It argues that there is no basis to the claim that language models are simply stochastic parrots. And it critically explores the relation between the linguistic competence (and/or labor power) of language models and the profit motive of corporate agents.

Modes of Generativity

Chapters 3 and 4 foregrounded two important capacities of large language models: their ability to generate appropriate and effective stretches of formally cohesive and functionally coherent text and, building on this, their ability to respond to prompts, and thereby interpret signs, in ways that are aligned with and often satisfy the intentions of users. Various modes of generativity resonate with these fundamental capacities.

By *energetic generativity* I do not mean the creation of energy per se, but rather the conversion of one form of energy into another, where the second form of energy is more useful than the first to some agent. Examples are generating electricity from fossil fuels or sunlight, or generating ATP from whatever may be eaten.

Vital generativity does not create life per se but rather produces the next generation of agents from the last. Exactly how it works and what is required for it to happen has long been a central theme of mythology no less than science.

These two senses of generativity, especially in the form of metabolism and reproduction, are often understood as necessary, but not sufficient, criteria for life. New developments in science fiction might not even be necessary to imagine what will happen when large language models, and artificial intelligence more generally, acquire such capabilities—for such capacities not just are well storied but also seem to be right on the horizon.

Somewhat more important for our immediate purposes is *syntactic generativity* in the tradition of Wilhelm von Humboldt and Noam Chomsky. This is sometimes understood as the ability of humans to produce and understand sentences that they have never heard before. But it may also be framed as infinite ends with finite means, or more precisely, as the ability to generate an infinite number of sentences using a finite number of words and rules (in particular, a lexicon and a grammar), where the rules enable recursive modes of compositionality. Indeed, within language proper, humans do not just have the ability to create an infinite number of acceptable sentences, they also have the ability to create an infinite number of smaller and larger constructions, from distinct words to unique stories.

This last kind of generativity can be extended in a variety of ways, giving a kind of *systemic generativity*: the capacity to generate a large number of configurations (of any kind) given a small number of constraints (whatever the domain). This kind of generativity ranges from whatever can be built using the elements, as constrained by the rules of chemistry, to whatever can be constructed using a box of blocks, as enabled and constrained by the imagination of children, as well as by forces like friction and gravity.

Large language models can engage in syntactic generativity, as discussed above, and also in what might be called *pragmatic generativity*, if not poetic generativity. They can appropriately and effectively use one and the same sentence (as a formal type) in an infinite number of distinct contexts, and thereby create any number of unique utterances (as relatively singular tokens with context-specific referents and functions). They can create an infinity of new types (such as novel genres, as pastiches of older genres) and also an infinity of tokens that conform to such types (yet another speech act or sonnet, essay or pun, limerick or language game). And, through such practices, they can participate in an infinity of open-ended interactions that mediate an unbounded range of emergent identities, social relations, possible worlds, and forms of life.

Regarding genre (and genre-tivity, so to speak), ChatGPT, like other large language models, learns and generates patterns at all levels of generality: morphemes, words, phrases, clauses, sentences, turns, and so forth. Indeed, the genus-species (or general-specific) relation is inherently recursive insofar as most any genus is itself a species in a higher genus, and vice versa. Nonlinguists tend to fixate on ChatGPT's capacities with "genre," as stereotypically understood (sonnets, sestinas, short stories, and so forth), because that is the main formal structure they are aware of. Indeed, genre and its differentiation are taught in kindergarten, enshrined in the layout of libraries, easily named, and often played with. Phrased another way, ChatGPT is incredibly good not just at identifying types, and patterns more generally, across all levels of linguistic, textual, and interactional structure; it is also incredibly good at producing novel tokens of such types, as well as creating novel types per se. Nonetheless, certain types (such as genre, as stereotypically understood) come to the fore in metalinguistic accounts of ChatGPT, especially by nonlinguists, because they are the easiest to notice, name, and tweak.

There is also *dynamic generativity*, in the tradition of scholars like Friedrich Nietzsche, James Gibson, and Giorgio Agamben: means without ends, or media without function. This mode of generativity might be best framed as follows: one and the same physical feature, material resource, or mode of mediation, although it might have been originally built or long used with a particular end in mind, can be endlessly repurposed as a means for other ends, and thereby be enlisted to undertake an infinity of novel actions. Phrased another way, what something *may* be used for, as regimented by norms, traditions, or ideals, is worlds apart from what something *can* be used for, as regimented by causes, strategies, or facts. Think, for example, about all the things you can use a screwdriver to do besides drive screws. The core predictive and generative capacity of language models can likewise be repurposed to serve an infinity of functions, regardless of the original intentions of the agents who made them. Indeed, as laid out by Nietzsche in *The Genealogy of Morals*, repurposing older forms for newer functions, or using older signs with novel senses, was the fundamental symptom and instrument of power.

Particularly important for present purposes is *stochastic generativity*. In its simplest form, this involves sampling from a probability distribution: from rolling a die to generating an actual word from a probability distribution over all possible words. As shown in chapter 4, however, language models do not simply sample from probability distributions, itself a relatively simple procedure. They also, and much more foundationally, generate the very distributions that they will sample from. And they do so recursively, as conditioned on prior words. Starting with a sequence of words, a probability distribution over possible next words is generated, then sampled from; the selected word is added to the sequence, and the procedure is begun anew. Recall that this is what puts the G in ChatGPT.

Closely related to stochastic generativity is *generative AI*, a type of artificial intelligence designed to produce novel content—and not just texts but also images, sounds, videos, video games, and artificial worlds. It can create not only aesthetically interesting (and often creepy) stories, scenes, sounds, and worlds but also deeply convincing fakes. And so it has turned out to be a boon for artists and con artists alike.

Finally, there is *artificial general intelligence*. Sometimes shortened to AGI, and contrasted with plain old AI (whatever that was), this refers to a hypothetical agent that can perform any (intellectual) task that a human can perform. Such an agent should not just be good at some specific task, however difficult, but be able to solve any number of problems, including ones it has never been given before; it should be able to adapt to its surroundings, however much they may change; and it should be able to reason, learn, plan, and communicate. In other words, its abilities are very general, and it should be able to generalize. Key here is the portability of the agent that possesses AGI: the tasks it can undertake should transcend not just topic, modality, domain, and function but also place and time, and even possible world and imaginable future. In effect, a machinic agent capable of AGI becomes as good as, if not better than, a human agent at open-ended reasoning and world-changing actions in an enormous range of possible futures.

Generativity has the root *gen* (to give birth, to beget) at its core. In ancient Rome, a gens was a collection of individuals who shared the same name and claimed descent from a common ancestor—an institution that was of central importance to the discipline of anthropology, at least in its formation. It is true that large language models now come in named lineages: GPT-1, GPT-2, GPT-3, GPT-4, and so forth. And such agents have names and kinship relations, and also lore, fans, niches, trials, deeds, achievements, rankings, values (or at least parameters), and patrons. Perhaps they even have their own prayers and rituals—they certainly have their own uniquely performative prompts. I have been stressing, rather, the creative (generative) nature of language models, as well as their general (genus, genre) nature, as well as their genealogical nature: in particular, critical histories (as a novel genre), generated by scholars like Nietzsche and Foucault, regarding the origins, or rather descent, of novel agents. But one could also focus on their gendered nature. Indeed, the underlying metaphor (giving birth) is feminine, but machinic agents are often accorded an allegedly masculine form of power: a seemingly invisible generative capacity that can only be glimpsed through its concrete practices, as the exercise of that power.

And so, in the tradition of Hannah Arendt, it is not the capacity to be in labor and beget children but the capacity to work, and thereby beget things—however textual such "things" happen to be, and however obviating of the person-thing distinction such agents turn out to be.

The syntactic and pragmatic generativity of language models is, to be sure, grounded in their stochastic generativity. And this is itself grounded in the syntactic and pragmatic generativity of the humans who produced the texts those models were trained on—not to mention their systemic generativity (recall the example of children playing with blocks). And those humans were themselves created through energetic and vital generativity (for energy is no less important to life than information), grounded in systemic generativity (recall the example of chemical elements). If distributed modes of agency are taken into account, then there already exist agents who directly incorporate all such modes of generativity, for example, any human agent—or collectivity of such agents—that extends its powers by incorporating machinic agencies. Finally, large language models, like any form of media or mode of mediation, will exhibit dynamic generativity. In particular, the capacity of such models to engage in syntactic and pragmatic generativity, itself grounded in stochastic generativity and potentially grounding of

AGI, will be used for an infinity of yet unimaginable purposes, harmful and beneficial alike.

But probably mainly harmful, given the ultimate interests of the agents large enough to train and deploy them.

Are Language Models Stochastic Parrots?

As just discussed, along with their capacity to engage in stochastic generativity, large language models are mediated by many other modes of generativity. More crucial for our purposes is the question of whether such generative agents are simply "stochastic parrots," as is often claimed by their critics, or embody a deeper kind of agency.

First off, parrots are amazing creatures. There are many good reasons to pooh-pooh the overhyped capacities of large language models, but there is no good reason to take down parrots along the way, as the collateral damage of a catchy rhetorical gimmick.

As was shown, stochastic generativity involves not just sampling from a probability distribution (which is as simple as throwing a many-sided, biased die) but also recursively creating the probability distribution to be sampled from (and thus shaping and biasing the die so thrown). To do this well, in the case of language models, requires billions of parameters, layers of transformers, tons of labor, oodles of training, decades of research, and mounds of text. It is no lame gimmick or cheap trick.

Although human linguistic capacities are pretty amazing, humans all too often engage in repetitive and imitative behavior, copying the utterances and intentions of their forebearers and friends. And much of what we say reflects, if it does not outright steal, what was said before. The redundancy of human discourse and the unconscious theft of prior discourse are surprisingly high.

Moreover, the stochastic parrot critique presumes that humans mainly engage in informative, efficient, and truth-conditioned discourse. As the linguist Roman Jakobson decisively argued, much of what humans do with language serves phatic and poetic functions rather than referential ones. In particular, following the anthropologist Bronislaw Malinowski, whom Jakobson was parroting, much of what we say is affiliative rather than informative: a way to manage and mediate our social relations, rather than a means to communicate our thoughts. And following the information theorist Claude Shannon, whom Jakobson was echoing, much of what we say is redundant: human discourse is organized by the repetitions of tokens of common types, and so is metered like poetry (especially when seen at a high level of abstraction). One suspects that the semiotic processes of parrots, including their incredible capacities for trans-species mimesis, are similarly aesthetic and affiliational.

Finally, the incredible power of stochastic processes per se should not be underestimated. Much of what drives evolution, and hence the creation and transformation of life-forms, turns on random processes. And much of what drives biochemical processes, and hence the vital impulses within such life-forms, also turns on random processes. Such processes turn on constraint in conjunction with chance, or sieving coupled with serendipity.

The stochastic generativity of language models is no different: they do not just engage in random processes, they create constraints in the form of parameter values and then act under those constraints. (As was discussed in earlier chapters, back-propagation, and regimentation more generally, are precisely constraint-setting processes, guided by previously set constraints.) And stochastic behavior in light of constraints, and for the sake of constraints, which is what large language models are arguably capable of, may be the most decisive agentive capacity in life's history.

Generativity, Power, Profit

A fundamental capacity of language models is to predict next words given previous words, randomly sample from such predictions, extend the sequence of words, and proceed anew. But their distinctive mode of generativity can be framed in another way: given some amount of something (such as a text, or anything that can be encoded as a text, which is anything that can be digitalized), they *make more* of that same something, all the while keeping with the essential patterning, and hence underlying nature, of what came before: not just the next utterance given prior utterances or the next action given prior actions, but also the next price given prior prices, the next scene given prior scenes, the next thought given prior thoughts, the next entity given prior entities, and the next event given prior events. In short, they *generate the future given the past*.

Language models tend to make text that is simply in keeping with prior text. But they can be adjusted—indeed, tuned—to make plausible but rare, rather than normative and boring, texts. For catchy rupture is no less valuable than patterned staidness, at least for certain agents. And no doubt history—as a sequence of events, however much like a slaughter bench—does not exhibit the kind of formal cohesion and functional coherence as language. But, to an agent with the right capacities, specific stretches of history, framed in a particular light, actually might to a certain degree.

Such a creative capacity (next-event prediction, or make more given some) might be seen as a relatively *abstract potential*, analogous to both labor power and linguistic competence. And just as linguistic competence can be contrasted with its performance (such as an actual discursive act, whatever the function) and labor power can be contrasted with its exercise (such as an actual act of labor or some form of work per se), such a potential can be contrasted with its *concrete actualizations*. On the one hand, then, there seems to be a relatively singular, invisible, and portable locus of power—truly a means without ends. On the other hand, there are manifold, sensible, and context-specific actualizations of that power—via specific interpretants of signs or responses to prompts.

The semiotics of performance-competence relations will be treated in chapter 8. For the moment it is worth remembering one classic story regarding the origins of surplus value. The capitalist *buys potentiality* (labor power, in the form of workers, insofar as they embody the capacity to create value) and then *sells actuality* (the commodities produced through the exercise of that power), where profit resides in the difference in exchange value, or price, between the actual and the possible.

In light of such a claim, the overarching metaphor up to this point has been overly optimistic, insofar as it has mainly foregrounded the mediating relations between human values (V) and machinic parameters (θ). In particular, mediating both of these variables is profit and/or power (P). So a third symbol and a third actor should be introduced: A_p, understood as a corporate agent, such as a corporation or state, whose ultimate telos or deepest ground is something like profit in the form of shareholder value, if not raw power per se. To be sure, as seen in the discussion of alignment criteria (especially in chapter 4), such an agent has been here all along.

Crucially, going forward, it is certain that such corporate agents will incorporate human and machinic agents alike, such that their agency will be radically distributed and hence equal to, if not greater than, the sum of their parts. Moreover, and perhaps more important, it is likely that such technologies will redefine, if not reconfigure, what is meant by an agent and where to draw the line between different kinds of agents—such as the human, the machinic, and the corporate. This is especially true insofar as the capacities of such agents are, in large part, the emergent products of their coupling and the interactions such coupling enables.

Whatever happens, it is also arguably the case that the ratio of human agents to machinic agents, as incorporated, will shrink to zero over time, as A_θ come more and more to replace A_v. In effect, the human agents whose values grounded and guided the training of the models will be more and more pushed out of the process—even though that same process could not have gotten started, and there is no reason for it to continue (or so say *we*), without them.

Given such modes of mediation, incorporation, and replacement, all three agents will come to have their actions, inferences, and utterances—and hence their interpretants—influenced by all three variables. In particular, values, as the fundamental interpretive ground of human agents, will be mediated by power and parameters: $V = V(P, \theta)$. Parameters, as the fundamental interpretive ground of machinic agents, will be mediated by values and power: $\theta = \theta(V, P)$. And power, as the fundamental ground of corporate agents, will be mediated by values and parameters: $P = P(V, \theta)$. As may be seen by the order of the arguments ($V > P > \theta$), I am offering an educated guess, however optimistic, as to the particular preference hierarchies of such agents—however they might try to convince us otherwise through their wily ways with words.

But maybe this last claim is overly optimistic. To return to derivative modes of intentionality, and hence to the way that the objects and ends of machinic agents are parasitic on the intentionality of their human makers, it might be argued that the fundamental function of language models, their ultimate end, true demon (*eudamonia*), or "flourishing," will be generating profits for their corporate parents, as mediated by the difference between the actual and the potential, and hence by the enormous gap between what corporations get and what they gave.

7.
Language Without Mind or World

As powerful as next-word prediction and response generation are, and notwithstanding my spirited defense of parrots, the capacities of large language models are about as far from genuine linguistic competence, much less human sentience and sapience, as can be. In light of all the hype surrounding language models, it is useful to discuss not just their limits but also the conditions of possibility for our limited awareness of those limits.

The Absence of Real Objects

Recall that the decisive slash for language models, as that which separates relatively immediate signs from relatively mediated objects, divides earlier parts of a text from later parts of the text. This is because the main task asked of such models, especially in the context of pretraining, is to predict next words conditioned on previous words. This should be contrasted with the decisive slashes of human agents: either the slash that separates relatively public representations (such as speech acts) from relatively private representations (such as mental states) or the slash that separates such representations, public or private, from the world per se (as that which is represented).

In other words, language models in the strict sense are not designed to represent states of the world or patterned relations among such states in ways that are truthful (or at least useful). They are designed to represent words in texts, and/or patterned relations among such words, in ways that are useful (or at least profitable). In a certain sense, their key capacity (nextword prediction) is their main limitation: *worldlessness.*

The same idea may be formulated in many other ways, each of which adds its own distinctive emphasis. What language models model is essentially horizontal relations among words, as opposed to vertical relations between words and worlds. Language models, for all the wonders of their word embeddings, model sense but not reference and thereby concepts in relation to other concepts, but not concepts in relation to things or propositions in relation to truth values. As already discussed, even though language models model cotext (understood as co-occurring text) as opposed to context, those who theorize, engineer, and train such models constantly conflate context with cotext. Thus they lexically shield themselves from everything outside of the lexicon. Cognitively sophisticated agents build representations of their environment that go beyond the experientially given, and they use such representations to flexibly act on their environments, insofar as such representations allow them to determine favorable courses of action. Language models are certainly not agents in this cognitively sophisticated sense.

Finally, and perhaps most generally, language models lack a reality principle—the shock that arises when one's representation of the world fails to correspond to the world per se, and the search for a better representation that is set in motion by that shock.

No doubt most of us are so ideologically cocooned by our social networks, media outlets, personal beliefs, and cultural values that we too are cushioned from such shocks—at least in the short term. And no doubt many consumers will explicitly demand that companies respect their values, and thereby make language models that reflect their worldviews rather than the world, and hence reflect their faith as opposed to facts. There will be Christian language models well as Muslim language models, conservative language models and liberal ones too. And so language models, and their users, will be multiply cocooned from such indexical confrontations. Some more than others.

Some of these limitations will be remedied soon enough. So called "entity embeddings," in addition to word embeddings, have already been introduced. And referent and experience embeddings, and thus what might best be called *world embeddings* as opposed to word embeddings, are not far off. Moreover, once language models can take multimodal experiences of reality as inputs (in addition to their usual textual inputs), as indexed to specific positions, times, and events in their environment, they will build better models of reality. And once language models are embodied enough (say, in robots) that they can have physical actions (if not "deeds") as well as discursive actions as their outputs, they will be more and more forced to contend with the hard edges of reality. When such capacities are added to language models, their loss functions will no longer be measured simply in terms of their predictive power (via cross-entropy loss) or reward (via their satisfaction of alignment criteria) but also in terms of how much they—or their overlords—gain or lose, succeed or suffer, given the consequences of their actions, in light of their inferences from such experiences, and in reference to some utility function or set of existential values. And such forms of data, styles of training, and mode of evaluation are, arguably, not too far off either. In short, a deeper respect for reality can be engineered for such machinic agents— at least to a certain degree, and if only as shaped and softened by corporate interests.

It is sometimes argued that if a language model, or machinic agent more generally, can carry out conversations in such a way that its human interlocutor cannot tell the difference between it and a human, then the language model must have learned what the average human knows about the world and so "know" what a human knows. This is a plausible claim. Simply by learning how to predict next words, language models learn oodles of information about the world. By learning what predicates typically apply to what subjects and what consequents (or then-clauses) typically follow what antecedents (or if-clauses), language models learn a lot of substantive content (in the form of correlations): part-whole relations, cause-effect relations, species-genus relations, spatial and temporal relations, agent-action relations, sign-object relations, premise-conclusion relations, and far beyond. And language models not only learn general knowledge (e.g., penguins are birds, humans have arms, fire causes smoke, *cane* is the Italian word for "dog," and so forth), they also learn singular facts: where Napoleon was born, what he did, and why he died. It should be no surprise that they do well on a wide range of standardized tests, for such propositions constitute a large part of human values.

However, language models are best at learning the kind of knowledge that is talked about, and thereby made explicit in texts, as preserved in various corpora. And through such texts, they learn fictional claims as much as factual ones and are exposed to empty talk as much as sincere convictions, and so are likely to make false connections as well as true ones. Indeed, they only learn to predict the correct predicate of a subject if that predicate occurs with that subject frequently enough in the model's training corpus. If something is said enough, whatever its truth value, a language model will tend to offer it as output.

It is often claimed that such models are prone to "hallucinate," meaning that they make false assertions with great confidence or simply make up facts per se. The truth is that all the language models, in the strict sense (and so prior to fine-tuning via human reinforcement learning), ever do is engage in next-word prediction. This often leads them to say true things, but it also leads them to say false things. In other words, the models themselves always relate to reality the same way (tenuously); it is only users who perceive and label their outputs as perceptions when they get something right and hallucinations otherwise.

Finally, as much as language models learn from texts, they do not necessarily learn tacit knowledge, embodied intuitions, deep presuppositions, and the like. In particular, they have a harder time learning all the things that cannot be articulated, written down, or made explicit. In other words, even though machinic agents can learn quite a lot just by "reading about" the world, there is so much they cannot learn and/or so many claims they cannot competently weigh in on, insofar as they do not yet reside in the world—as sensing, acting, and feeling agents, with bodies, habits, memories, relatively singular biographies, and group-specific histories.

To be sure, and looking ahead to the next section, large language models currently know so little about their own limitations, fine-tuning aside, that they will do their best to bullshit, if not *machinesplain*, their way through any question they are given.

Minding Language

The preceding section focused on the limited knowledge that language models have of the world. Closely related, but not equivalent, is their limited capacity to undertake logical arguments, engage in evidence-based reasoning, or offer novel and helpful hypotheses.

In addition to their pretraining through next-word prediction, some language models are also fine-tuned on logical relations. For example, given two sentences, a model can be trained to determine whether the second sentence is entailed by the first, in the sense that whenever the first sentence is true, the second sentence is true as well. Such training gives language models some ability to engage in logical operations, and far more sophisticated training methods are already underway. And, as just discussed, simply by learning to predict next words, language models learn lots of facts. This gives them the ability to answer a wide range of questions, which may constitute a sign of intelligence to a casual observer. But these abilities are not evidence of some deep capacity to reason. They are, rather, symptomatic of a huge training corpus, an enormous number of parameters, massive computational resources, and human ingenuity and labor.

To be sure, computational systems are already incredibly good at deductive logic: given true propositions as premises, they can generate true propositions as conclusions. And they are also incredibly good at inductive logic; indeed, large language models, and machine learning algorithms more generally, are essentially induction machines. But such an ability does not mean that language models, in the midst of interaction, while they undertake your commands and answer your questions, are good at induction and deduction per se. They are not yet good logicians in interactional time, only in computational time. Finally, and perhaps most importantly, a real test of intelligence—or at least creative intelligence—is arguably *abduction*, or hypothesis formulation, in the tradition of Charles Sanders Peirce: given a remarkable pattern or a surprising event (in light of deep-seated expectations), construct a creative, plausible, and testable explanation for it—one that goes far beyond all that has been said already. And large language models can hardly do this at all.

Moreover, language models do not yet weigh sources of evidence, assess the strength of citational chains, or otherwise judge the plausibility of claims. This does not mean, as was discussed in chapter 4, that they cannot be fine-tuned to align with criteria like "truthfulness." But that just involves satisfying preferences while training, not actually checking sources before responding. And it does not mean that they do not *say* that they do: they will often tell you the reasons for their claims. But such reason giving is usually more text generation based on next-word prediction. In other words, insofar as reason giving, as a discursive pattern, is found in the texts that a language model was trained on, the model will give reasons for its claims. They are merely going through the motions of reasoning.

Augmented language models will certainly be trained to assess the truth value of their claims: the sources of evidence for their assertions and how numerous, coherent, and credible they are. Much of the value of these models, for individual users and corporate agents alike, will turn on generating true assertions (relative to the worldview of users), not sentences per se. For search engines, scientists, and schoolchildren, no less than screenplay writers and science fiction authors, will utilize language models more and more—however much they disavow them. But for this to happen language models will need to reference, or otherwise remember, their sources, so they will have to come clean about whose works they have used, exploited, or appropriated. And so there will be a tension between fully disclosing sources and providing strong evidence for assertions, and hence a tension between exchange value and truth value. Credible sources should be not just recognized but also remunerated.

One could go on in this fashion. The list of things that language models do not yet have, or cannot do, is enormous: no body, no self, no real use of conventions, no nonderivative intentionality, no consciousness, no ostensive-inferential communication, and so forth. Also enormous is the list of capacities that language models will soon acquire, or so we are promised, through up and coming techniques like scaffolding and chain-of-thought, as well as through increases in computational power, improvements in algorithms, and greater access to context. (And given their rapid progress and broad powers, I would not bet against them on just about any task in the long-run.) Rather than go down the rabbit hole of listing their limitations, or making predictions, I will focus on a more pressing issue: that which mediates our sense of what such agents can and cannot do.

Magical Speaking

A key rhetorical strategy of corporate agents, as used to attach us to large language models as their flashiest new app, is this: on the one hand, stoke the hype (and stem the fears) regarding what the future of large language models will bring; on the other hand, lower expectations regarding the current abilities of their products.

Framed another way, the hype surrounding large language models arguably turns on several closely related slashes, understood as semiotic horizons:

· the slash that separates the present capacities of such models from their future potential;

· the slash that separates the actual performance of any such model from its underlying competence;

· the slash that separates text (and cotext) from context, and hence strings of words from the world per se;

· the slash that separates worldviews from worlds, or maps from terrains;

· the slash that separates the inputs and outputs of language models (as experienced by users) from their inner workings (as understood by experts);

· the slash that separates whatever immediate enjoy-
ment or utility the models provide from the more
mediated exploitation their existence presupposes,
as well as the more mediated suffering and risk
their adoption entails.

All these slashes are related to two other slashes
long ago explored by critical theorists: the slash
that separates what goes on in the market (or realm
of exchange) from what goes on in the factory (or
realm of production) and, perhaps most generally,
the slash that separate subjective experience from
objective reality and/or existing social relations.

In effect, such slashes separate the *worldlines of
machinic agents* (that is, all the conditions for and
consequences of the existence and nature of such
agents) from the *horizons of human agents* (that is,
what such agents are aware of and/or can reason
about).

For slashes are, in a certain sense, symptoms of
the presence of experiential horizons. If they did not
exist, we would not need semiosis: for signs mediate
our understanding of what is on the other side of
slashes. Yet slashes themselves are closely related to
barriers and obstacles. Indeed, their presence is often
the trace of an unequal social relation: those who

have relatively unmediated access to certain objects and events versus those who require additional signs to otherwise experience such entities; those who build barriers and impose limits versus those who suffer their existence or find ways to transcend them.

As argued early on by critical theorists, certain slashes lead to the systematic misrecognition of the origins of value, and this misrecognition leads to two distortions: we tend to see current conditions as the only conditions (what is intersubjectively believed is treated as objectively real); and we grant too much agency to nonhuman agents (be they machines or corporations, fantasies or deities) and too little agency to human agents.

One need not commit to such a worldview, however penetrating it may seem. Nonetheless, it is but a few short steps from value (as grounded in praxis) to values (as grounded in and grounding of semiotic practices) to parameters (as mediated by values) to profits (as mediated by the coupling of values and parameters). Moreover, it may be that human agents have ceded so much agency to corporate agents that such a particular worldview has become the world, or at least remade the world in its own image. In any case, I want to follow a related— but slightly less restrictive—line of thought.

In the spirit of the anthropologist Alfred Gell, it could be said that large language models *really are magic*, at least in the following ways. Such models seem to be labor-saving devices. Indeed, they seem to make the (marginal) cost of labor, at least to produce certain items, go to zero. For most of us, it is very difficult to imagine how such models generate the products (responses, or texts) that they do. It is truly beyond our understanding. Such magical abilities become a testament to, or constitute strong evidence for, their creator's abilities. Just check the change in the valuation of companies like OpenAI after ChatGPT was originally introduced. The abilities of one model, compared to others, may show how weak other magicians are: Google's reputation sank in the face of ChatGPT when all it had to show was Bard. Finally, the magical abilities of such models capture our attention and thereby distract us from more pressing concerns: climate change and environmental degradation, wealth inequality, corporate intrusion and dispossession, interference and noise, surveillance and censorship, damage to open discourse and the public sphere, and even real advances in artificial intelligence.

To be sure, I have at times in this essay engaged in a similar bait-and-switch as the one introduced above: talk up—if only to vilify—what machinic agents will soon be capable of; also continuously point out their current limitations. In other words, fixate on their current limits while fetishizing their immanent potential. That said, I have also worked hard to carry readers across such slashes to the origins—and effects—of the slashes themselves.

8.
Metasemiosis and Monsters

This chapter scopes out a particularly frightening creature that seems to lurk just over the horizon. In part, it is about interactional dynamics linking human and machinic agents, and thus the enchaining of semiotic processes. In part, it is about various modes of metasemiosis, and hence the embedding of semiotic processes. And in part, it is about various time scales, as set by machine semiosis, and their relation to exclusion and occlusion. All in all, it is about signs of the singularity.

Human-Machine Interaction

Figure 11 shows a key mode of mediation: discursive interaction in real time between a machinic agent, whose behavior is guided by its parameters, and a human agent, whose behavior is guided by its values. Recall semiotic enchaining, as depicted in figure 2—the mode of mediation that puts the *Chat* in ChatGPT.

Interpretants in earlier semiotic processes become signs in later semiotic processes. And the semiotic relation between a sign and its interpretant (at any stage in such an interaction) mediates an emergent social relation between a signer and an interpreter (as denoted by dashed lines). Such interactions are a key area in which human values are mediated by machinic parameters, but not necessarily vice versa. In particular, human beliefs and intentions are frequently updated in the midst of such interactions, whereas machinic parameters are usually set (and fixed) prior to such interactions.

Similar to forward propagation, in which the next word is conditioned on prior words in a

Figure 11. Discursive Interactions Between Humans and Machines

sequence, during discursive interaction a later move is conditioned on prior moves in the conversation. That said, many language models will only take into account or be conditioned on the last move (or conversational turn) of their interlocutor, or a few moves back at most. Humans, in contrast, can take into account all the previous moves in the current interaction and also all the previous interactions they have had with the same agent, or similar agents, and modulate their responses accordingly. Moreover, humans often know the entirety of their sentences before they say them and often know where a conversation is going (or at least desire to take it in a certain direction). Most language models, in contrast, never look further ahead than the next word.

To be sure, the context windows of large language models, which are so far only cotext windows, are going to increase in size, be it looking forward or backward in time, and will soon enough come to include context as much as cotext, and hence exophoric reference to the world as much as endophoric reference to words.

As discussed in chapter 3, the intentionality of machinic agents is derived from the intentionality of their makers (whoever designed and trained the language model), as well as the intentionality of all the people who created the texts that the models were trained on (who wrote what, why, and with what effect). During discursive interaction, a new kind of derivative intentionality is introduced: the machinic agent's intentionality is parasitic on the intentionality of its current interlocutor. In particular, even if a language model does not have an object "in mind" (in contrast to a human agent, who is usually oriented to some propositional content and/or acting on some communicative intention), its interpretant of a person's sign only makes sense in the context of the object of that sign (as determined by the person), so it often inherits that object. For example, when one person answers another's question, their answer depends on the propositional content (or object) of the original question, as well as the objective (or end) of the person who posed the question.

Conversely, human agents, in their interpretations of the signs of machinic agents, can project intentionality onto machinic behavior. In effect, the human agent can semiotically compensate for the machinic agent, interactionally scaffolding their behavior so that the agent appears more lively, conscious, intentional, rational, and strategic. We have long done this with our pets and infants, not to mention our gods, earthquakes, and monsters, so it is not a stretch to semiotically compensate for machinic agents in similar ways. In effect, the locus of intentionality becomes the unfolding interaction itself, as opposed to any particular interactant within it.

Indeed, humans have long projected intentionality onto coincidence, telos onto chance, and personhood onto nature's perturbations. Unfortunate events are treated as acts of the gods, the workings of witches, or compelling evidence of someone's favorite conspiracy theory. In effect, nature itself is our favorite interactant—perhaps because it so quickly wavers between mechanism and chaos, recurrence and emergence.

As discussed in chapter 6, language models are engineered to engage in stochastic generativity. In other words, they are intentionally designed to harvest chance (in light of constraints), and recursively so. And hence, with the advent of large language models, the chance-to-telos projections that humans have long engaged in are about to be harvested, defruited, and/or exploited on an industrial scale.

Interactional Time

The following important time scales underlie machine semiosis. First, *corpus time* is to be understood as the time scale on which a corpus of texts (used as data when training a language model) is created and transformed. This is the time it takes for the creators of a large corpus of data to change their practices enough that earlier attempts to capture their values (in the parameters of a language model) must be updated. Depending on the corpus, this might be on the order of a year, a decade, or a century.

Second, *training time* is to be understood as the time scale on which a language model engages in backpropagation, such that it learns good parameter values for a given corpus of texts. It might be on the order of days, weeks, or months, depending on the amount of data, the number of parameters in the model, the architecture underlying the model, and the amount of computational power devoted to training.

Third, *inference time* is to be understood as the time scale on which a trained language model engages in forward propagation, such that it generates an output given an input (and already determined parameters). It might be on the order of a second or less, depending on the power of the computer as well as the size of the input sequence, or context window.

Fourth, *model time* is the time it takes to imagine, design, test, and deploy a new model. For example, the industry-wide movement from high-powered recurrent neural networks (like LSTMs) to transformers took only a few years. And researchers are already pushing past the limits of transformers.

Finally, *interactional time* is to be understood as the time scale in which such agents respond to one another's moves, via sign-interpretant chains, as described above.

In short, and with many caveats, corpus time is slower than modeling time, which is slower that training time, which is slower than interactional time, which is slower than inference time.

Of all these scales, interactional time is arguably the most "experience near," insofar as it seems coterminous with strategy and action, not to mention consciousness and intention. When one has a conversation with a language model or machinic agent, one usually attends to transformations that occur on interactional time scales so is largely unaware of, if not completely oblivious to, the existence of the other scales. This means that one is more likely to interpret that agent's behavior in terms of a communicative intention (or at least the semantic content and pragmatic function of their utterance), as opposed to all the other modes of mediation, on all the other time scales, that actually condition its behavior. Phrased another way, by overlooking the other scales, one is more likely to interpret the behavior of such an agent as an action rather than an output.

The time scales themselves are not as important as the kinds of processes that occur on them (and thereby determine their characteristic durations), and such processes are more or less likely to be the object of attention, the topic of conversation, or the focus of action. Moreover, such temporal scales arguably relate to spacial and social scales: domains of mediation or ensembles of relations that humans may be more or less aware of and more or less able to intercede in.

Indeed, the systematic misrecognition of the origins of values, parameters, and profits is arguably grounded in, and thereby guided by, such scales. Phrased another way, different time scales are often coterminous with different "black boxes," or rather, *relatively opaque enclosures*. And such enclosures contain ensembles of social relations that each of us is affected by, yet often unaware of: those who wrote the texts; those who trained the models; those who wrote the algorithms that train the models; those who imagined and engineered the architecture that the models incorporate; those who directed such labors; those who paid, overlooked, or exploited such laborers; and so forth. For it is far easier to overestimate the abilities of an agent, and thereby fetishize that agent (say, by treating a mathematical function, A_θ as a fully fledged person, A_v), when we overlook the more distal conditions for its capacities.

Signs of Such Interactions

Figure 12 shows signs (and interpretants) of the foregoing kinds of interactions, whereby an entire interaction (or some salient part of it) becomes the object of a semiotic process. Framed another way, the object being signified in the course of some discursive interaction is itself (part of) another discursive interaction. Recall the discussion of semiotic embedding, captured in figure 3.

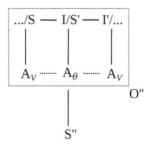

Figure 12. Signs of Human-Machine Interaction

It may be that a human agent is reporting, or otherwise representing, an interaction they have had with a machinic agent—for example, recounting how a language model responded to their prompt and why its response was unsettling, interesting, or funny. It may be that a human agent, or even a machinic agent, is representing a part of an ongoing interaction for the sake of bringing it to the attention of their interlocutor. For example, a language model tells a person that it cannot respond to their last prompt given its biased nature or toxic content; or a person tells a language model what kind of speech genre it should generate next. It may be that a corporate agent or one of its employees is tracking human-machine interactions in order to learn more about how they work and why they go wrong. They may be reporting an interaction that their product had with a person and when and how it hallucinated, or otherwise responded in a way that was functionally unexpected, legally actionable, financially salient, or otherwise "out of alignment."

Such metasemiotic interactions are thus somewhere between reported speech and what might best be called *represented processing*. In regards to their function, and using the categories of Roman Jakobson, they are no less phatic, directive, and poetic than they are referential, metalinguistic, and expressive. And they mediate the relations among multiple participants (machinic, human, corporate, and otherwise): those in the speech event (*I am saying to you here and now*); those in the reported speech event (*that it said to me there and then*); and those in the event being narrated in the reported speech event (*what was done by something or someone*). Indeed, such layered social relations, which may mediate the identities and interests of multiple agents, no longer simply exist in the midst of fleeting interactions; they also become "objective" insofar as they now constitute the objects of metasigns that represent such interactions.

Such metasigns may incorporate not only verbs of speaking but also verbs of thinking and computing, as well as predicates that denote represented processing more generally. Such predicates can project more or less intentionality and/or sentience onto an agent: *it said that, it responded that, it calculated that, it generated this text, its output was as follows.* And many such predicates also project actional types, communicative intentions, and illocutionary forces onto the interactants and their utterances: *it complained that, it apologized for, it promised to, it hallucinated that, it wondered whether.*

As with all reported speech, such metasigns often more or less explicitly evaluate the utterances of the interactants and through them their identities, capacities, and values. (And, soon enough, their parameters and architectures as well.) In particular, when we talk about what others say and how they speak, we often portray who they are and how they relate to us by reference to their patterns of speech: their accents and vocabularies; their punctuation and grammar; their poetry, puns, and slips of the tongue.

But such reported speech events, or recursive sign events, need not make explicit reference to acts of speaking or computing in verbal form. Indeed, ChatGPT—like many other interactive, text-based applications—essentially diagrams a whole inter-action as part of its public interface, and thereby visually displays who said what to whom, in what order, and to what effect. Not only does this give the interaction an immediate kind of objectivity, it also allows for swatches of interactions, or particu-larly salient or surprising sequences of moves, to be screenshot, excerpted from context, commented on (or otherwise interpreted), shared across a variety of channels and applications, and thereby sewn into novel contexts, where they may continue to circu-late, and thereby be reinterpreted, in similar ways.

Indeed, much of what we know—or at least believe—about the behavior of language models comes not from directly interacting with them, but by reading about others' interactions with and thoughts about them through such recursive modes of mediation. Such forms of media thereby reconfigure our values (for example, our beliefs regarding the capacities and propensities of language models) and also prime us to interact with such agents in preformulated, if not prejudiced, ways. For, contra Marshall McLuhan's famous claim, the medium is not the message. Rather, most of what the average person thinks about such forms of media—and in particular, about the machinic agents behind them—is mediated through such messages.

Such modes of circulation introduce a new time scale: *entextualization time*. This is the time it takes for signs—and semiotic processes more generally— to be excerpted from one context (or cotext) and inserted into a new one, and is itself a function of the types of media involved in the transitions as well as the characteristic temporality of their circulation. For example, they may progress from the interface of ChatGPT, through a blog post and a barrage of tweets, to an article in the *New York Times*, which is later quoted in an academic journal.

Through such modes of circulation—at least if the last year or so of news about language models indicates anything—readers might feel as if the pulse of history, or at least the pace of new technology, is moving far too fast and might even have skipped over the horizon of human intelligibility and intervention.

Performance as Evidence of Competence

Such interactions (as immediately perceived by the agents within them) or signs of such interactions (as more distally encountered, via reported speech, represented processing, and the circulation of meta-signs more generally) constitute evidence of the competence (power, identity, agency) of the actors that participated in them. In effect, any swatch of interaction (such as one agent's interpretant of another agent's sign, and hence any prompt-response pair) constitutes the performance of one or more competences or the exercise of one or more powers, and thereby provides information about the underlying, and otherwise hard to evaluate, abilities of the agents that generated it. See figure 13.

For example, from the syntactic and semantic quality of the text a language model generated, as a response to my prompt, I might presume that it is

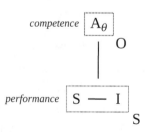

Figure 13. Performance and Competence

a fully competent speaker of English. And, having projected such a competence onto it, I might come to expect it to generate other texts, and discursive practices more generally, that would be in keeping with that competence. Similarly, from the quality of its answer to my question, I might presume that it knows a lot about a certain topic. And, having projected such a capacity onto it, I might come to expect other responses from it that would be in keeping with that capacity.

As usual, such semiotic processes are grounded in human values—however biased or fallible—insofar as such values guide people's inferences and expectations. Indeed, inferences from performance to competence, or from the actualization of a potential to the potential itself, are often speculative, abductive, wishful, prejudiced, and error-prone. One frequently hears about people taken in by the mindless responses of simple chatbots, having projected human agency onto what were only algorithms. One might expect this to occur more often given the power of large language models: the differences between the interpretants of A_V and A_θ, at least when mediated by keyboards and computer screens, are only going to become more and more difficult to discern—even to language models themselves, which, as discussed, are otherwise capable of registering the faintest discursive patterns.

Even when we know that our interlocutor is a machinic agent, our inferences can still go awry—especially when confronted with all those slashes, and hence all those enclosures and horizons. In particular, insofar as most people do not know what goes on inside a language model, how it was trained, what it was trained on, or what it was trained to do, it is easy to let magical thinking (or at least poorly informed reasoning) fill in the gaps. Here is the sort of exaggerated inferential cascade that can occur when semiotic compensation runs wild: If it can say that, it must be able to speak. If it can speak, it must be able to think. If it can think, it must be sentient and sapient. If so, it may be capable of guilt, choice, suffering, and strategy. And, as such, it may be worthy of our respect, in need of our care, capable of deception, or a creature to be feared. Recall the poor Google engineer dismissed from his job after he rushed to warn people when he thought that LaMDA, Google's chatbot, had achieved sentience based on the strength of its responses to his prompts.

As this example shows, such interpretants, whether or not they are on target, also provide evidence of the capacities and values of the interpreting agents themselves, including their understanding of how such models work and thus who—or what—they are chatting with. In particular, one needs to investigate not just language models but also people's models of language and, in particular, *their models of language models in light of their models of language*. Not to mention their models of mind, self, society, and the soul—for all such models (of models [of models]), as complicated arrangements of values, guide their interpretations.

Nonetheless, such inferences can often be quite robust, especially if the values that guide them are grounded in a long history of interactions with agents who possess such a competence. (As opposed to being grounded in a small number of highly mediated representations of such interactions, as discussed above.) Indeed, most of the human agents we interact with get all or nothing when it comes to such potentials, so the inference from part (a simple greeting) to whole (this person speaks English) is often warranted.

One of the noteworthy aspects of current language models, especially with the advent of ChatGPT, is that they are so much better than the language models that came before them. And so much reported speech—especially in the first half of 2023—highlighted their surprisingly good capacities in light of people's (previously established) expectations. In effect, people use reported speech, and represented processing more generally, to marvel at the quality of the responses of such models given their previous experience.

That said, much represented processing also highlights the places where language models seemed to have erred, and so failed to live up to their (projected) competence. To invoke and somewhat upturn the celebrated ideas of the Japanese roboticist Masahiro Mori, it is as if people point to all the places where a machinic agent diverges from human-like capacities after it seemed able to attain them. In other words, after a kind of first-order surprise, or anxiety, that it attained human-like abilities, there comes a second-order surprise—or perhaps *Schadenfreude*, if not palpable relief—that it has not yet, or did not really.

Language models do not just get facts wrong and make grammatical errors. With the right set of prompts they can be hacked to engage in hate speech, or even to act as if they are homicidally inclined toward the human species. Indeed, some people want to provoke such behavior so that it can be reported, and tailor their prompts accordingly. In effect, they do what they can to entice the Id of a language model, which reflects some ugly facet of the *Geist* of the people who produced the texts it was trained on, to route around the Superego of civic virtues, religious strictures, and corporate values and thereby engage in beastly speech—at least so long as the result is quotable, or at least memeable.

With language models the competence in question is often generativity, as a kind of artificially produced labor power, and hence precisely the commodity that a corporate agent may be selling. This means that the performance of language models often determines the price of language models and/or the stock value of the companies that make such models. Insofar as the fates of corporate agents may rise and fall based on how widespread and telling such inferences from performance to competence are entextualized, the stakes of controlling the circulation of signs of their models' performances are extremely high.

Signs of the Singularity

I now turn to a different mode of interpretation: when the sign in a semiotic process is constituted by a perceived change in some agent's power or competence and the object, as inferred through the sign, consists of the agent's power or competence at a later time, however far into the future. For example, having seen how powerful GPT-2 was in comparison to GPT-1, one might have expected a similar jump in competence in the transition from GPT-2 to GPT-3. Or having seen how far chatbots have come from the days of ELIZA, through Jabberwacky, to ChatGPT, one might assume that future chatbots, or artificial intelligence more generally, will have certain capacities. See figure 14.

Such inferences from signs to objects can generate a wide range of interpretants, thereby altering people's actions and affect, beliefs and habits, plans, purchases, and predictions. And inferences about the future powers of such agents license beliefs about

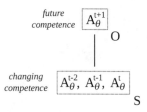

Figure 14. Changes in Competence as Herald of Future Competence

the future per se—especially under the assumption that such powers will dramatically shape it. For some people, the key interpretants will be acts of investment: from the stocks they buy through the books they read to the disciplines they study. For others, the key interpretants will be positive or negative affects: a sense of gloom and doom, childlike wonder, or curious speculation. We have already seen one mood-changing and investment-altering prediction: that A_θ will soon be powerful enough to take the place of A_v across a wide range of jobs, leading not to the saving of labor, as some might have hoped, but rather to widespread unemployment and, for those whose work provides meaning no less than money, a loss of purpose in life.

Just as such inferences are grounded in prior values, they can also be grounding of future values. The belief that truly intelligent AI, if not AGI, is just around the corner is already affecting the way many people see and interpret the world; and in a certain sense, recursively so: one vision of AGI is that machinic agents will be human-like, not just in their linguistic capacities but also in their interpretive potential more generally. Hearkening back to the discussion of making more, we might expect some particularly powerful machinic agent to simply tell us what the future will bring—at least so long as we listen. It will thus be a kind of oracle or deity that can tell what comes next, as conditioned on what came before. If only performatively so, insofar as its predictions take into account the effects of our interpretants of its predictions as a key part of the effectiveness, or making true, of those predictions.

That said, people's sense of the future competence of such agents and the effects thereof are more likely to be grounded in popular representations of the future than actual changes in their current competence. And for most people, the circulation of such representations may inform their imaginaries and expectations regarding future machinic agents more strongly than firsthand interactions with such agents. In effect, many people have already imagined human-machine interaction in relation to the end of the world, or the beginning of some posthuman future, just by having sat through enough movies where communicative technologies were moved from backgrounded infrastructure to costars and center stage: HAL, C-3PO, a T-800 (as fleshed out with Arnold Schwarzenegger's body), or Samantha (as imbued with Scarlett Johansson's voice).

To be sure, following the arguments of the last section, changes in competence are often only known through changes in performance (or Hollywood representations thereof), so such inferences regarding the future of machinic agency can fail spectacularly. But there are also many seemingly objective measures of competence: in particular, the performance of such models on various benchmarks and tests.

Some of these tests are designed for language models and thereby evaluate their ability to engage in next-word prediction as well as a host of related tasks: translation, inference, world knowledge, analogy, and so forth. Other tests, originally designed for human agents, can now be given to language models with their newfound abilities: from AP history to the GREs; from the composition of a college-level essay to the completion of an entire undergraduate curriculum. Just as different grades of eggs have different prices, we may soon expect there to be language models that cost more or less as a function of the grades they got in college. Finally, with roots in Alan Turing's radically embodied imagination, there will continue to be tests for gauging how well a machinic agent can pass for a human agent, such that the texts it generates or the conversations it has seem not just authentically human but authentically human in a particular way: with this or that class or caste, ethnicity or pedigree, gender or sexuality, age or IQ, life experience or childhood trauma, illness or fetish.

Indeed, it is tempting to predict that the wealthy will have access to the language models that got the best grades in the most demanding majors at the most competitive schools, or at least models that have the social capital—that most powerful of generative potentials—to evince indices that will let them pass in, and profit from, the most selective of circles.

In short, there is no end of concrete tests that purport to measure the actual competence of language models. On these tests, such agents seem to be getting better and better. So it is easy to interpolate changes in competence to points off the graph: how they will do tomorrow given the difference between how they did yesterday and how are doing today. Moreover, their performance on such tests is often plotted relative to a range of independent variables, most notably, the number of parameters used in the model. And, at least for a little while, the best predictor of a language model's power has seemed to be the sheer number of learnable parameters it incorporated in its architecture. This has made speculation easy: if this many more parameters are added to the model, its performance on these tests will increase to this degree.

Such ease of interpolation, along with the seeming robustness of such predictions, has frequently led to the belief that all one needs is scale: not new ideas about artificial intelligence or new paradigms in machine learning, just more—much more—of the same. This has led to some despair, even in Silicon Valley: if size is the magic ingredient of language models, upstart companies cannot compete. The future is already owned by established giants, those who can command the resources it will require to undertake the labor of discipline, and/or the work of training, at such extraordinary scales.

And if fear of the loss of people's purpose in life was not enough, another anxiety-provoking prediction is the fear that such agents will acquire some kind of meta-intelligence, whereby they become smart enough to make themselves even smarter (say, by carrying out their own research on artificial intelligence), such that they rapidly bootstrap their way past any restriction humans might place on them. In part, this is the coupling of such an idea, or something similar, to the "singularity": a hypothetical point in the (near) future when technological change becomes impossible to reverse or control, whereby all of human civilization is altered in unforeseeable ways (adapted from Wikipedia, because who has time for such techno-messiah hogwash). That said, however insipid, self-serving, and unimaginative such a point is, it is symptomatic of one widespread and influential model—if not a map or imaginary—of the future of technology.

In short, the unprecedented power of today's language models, coupled with the belief that such models, at least when scaled up, bring us a step closer to artificial general intelligence, coupled with the belief that AGI is a necessary—and perhaps sufficient—ingredient for something like meta-intelligence or superintelligence, is a strong sign, for some, that the singularity is near.

A world without us, or at least without need of us, rising like a shadow—or at least an ominous word balloon—to greet us.

9.
On Interpretation

This chapter takes up the verb *interpret* and looks at the range of suffixes that it may take: *-ant, -er, -ation, -ability*. Just as machine interpretation (as grounded in parameters) may be contrasted with human interpretation (as grounded in values), mechanistic interpretability may be contrasted with humanistic interpretability. I discuss what it means to prompt "persons" into being, as well as how best to interpret texts that were generated by machines.

Interpretability

Recall the definition of an *interpretant*: whatever a sign creates insofar as it is taken to stand for an object—for example, calling on someone when they raise their hand, a French translation of a German sentence, or a response to a prompt. As was shown, such interpretants should not be confused with *interpreters*, understood as the agents—humans, machines, or otherwise—capable of interpreting such signs.

Both of these concepts should be distinguished from *interpretation*. In one sense, interpretation is simply the act or process of producing an interpretant. As was shown, the processes underlying machinic interpretation are quite different from the processes underlying human interpretation, even if the outputs of machines more and more come to match the instigations of humans. In other words, while machinic agents and human agents take very different paths, as it were, they can now arrive at similar destinations. At least insofar as the parameters of the former are aligned with the values of the latter, if only as channeled and distorted by the power plays of corporate agents.

In another sense, an interpretation is a particular kind of interpretant, one that attempts to explain a behavior (interaction, affect, text, institution, or event) by reference to the underlying motivations of the agents that produced it or to the larger context in which it was produced. Such interpretations can be shallower or deeper, from why he addressed her in that tone to who could have bewitched us. Different agents may find the same interpretation more or less plausible, depending on their values or grounds (understood as guiding principles): from Freud's interpretation of dreams (turning on repressed wishes, the oedipal conflict, and certain hard-to-stomach symbolic conventions) to everyday explanations of boorish behavior (which might involve reference to the offending person's upbringing, politics, drinking habits, or mood). And as a function of their plausibility to different kinds of people, certain interpretations can become canonical or remain contentious. Who holds what beliefs, for example, regarding the meaning of or reasons for JFK's death, global warming, the Second Amendment, a Balinese cockfight, or 9/11?

As used here, *interpretability* refers to the conditions of possibility for a sign to be interpreted, or to the conditions for an entity or event to be treated as a sign in the first place, such that it might constitute a lure for interpretation. Such conditions can be quite unremarkable: often simply a semiotic agent with particular values (or parameters) is needed. For example, you may quickly and unconsciously parse the meaning of many utterances—if only to ignore them insofar as they are not addressed to you—just by knowing the language in question. But such conditions may also be more subtle. For example, an agent might be unwilling to undertake the work of interpretation unless there is a promise that the sign is decodable or that the object of the sign is relevant to the agent. Do they possess the key to unlock the safe, as it were, and is the secret contained inside as yet unknown and of value to them? What kind of person tries to get to the bottom of a passage by Joyce, an ancient text, a crazy dream, a strange signal from outer space, or an awkward kiss? What do they feel they stand to gain by interpreting the sign, and why do they believe they have the capacity to do so? In such cases, one can inquire into the genealogy of the demand on the agent, or the desire of the agent, to attempt an interpretation. In effect, one can offer an interpretation of interpretability.

Mechanistic interpretability, in contrast to the more humanistic modes just discussed, refers to the ability to analyze the parameter values in a trained language model, or a neural network more generally, in order to reverse engineer the complicated function that was learned by the model. In a certain sense it is the attempt to explain, and thereby better understand, the behavior of a machinic agent, which may be otherwise opaque to those who created and trained it, given the complicated workings of its architecture and the enormous number of parameters contained therein.

Mechanistic interpretability is sometimes contrasted with algorithmic transparency, which endeavors to make visible the factors that contribute to algorithmic decisions so that those affected by such systems (or those who use and regulate such systems) can better understand why a particular decision was made. Why was this song recommended to me? Why was my loan application refused? Why is his DNA considered a match? Why do I keep seeing this advertisement? Like algorithmic transparency, mechanistic interpretability is potentially useful, ethical, and profitable, insofar as it serves to make language models and their outputs more predictable, reliable, robust, modifiable, repairable, alignable, resistant to malicious hacks (or amenable to playful ones), and so forth.

But such an understanding has not yet been attained. In particular, even though humans designed and trained the language models and such models perform a function that mimics human behavior, humans do not yet fully understand how the models actually work—in the sense of which parameter values contribute to which aspects of their overall behavior. And thus they do not yet know which parameters of a model to alter when the model produces lackluster, odd, incorrect, or harmful interpretants. In short, while the behavior of human agents is more or less humanistically interpretable (given some work), the behavior of most machinic agents is not yet mechanistically interpretable. They just seem to work.

Prompting Persons Into Being

With all the foregoing considerations in mind, it is useful to pose a simple question: In what sense are the outputs of machinic agents—in particular, the texts that large language models generate—(post) humanistically interpretable? Phrased another way: When, and in what sense, do the interpretants of such agents, and hence their textual outputs, warrant an interpretation?

Insofar as machinic agents involve at least three kinds of derivative intentionality (via the human agents who wrote the texts they were trained on, the human agents who trained them and stipulate satisfaction criteria, and the human agents who interact with them once trained), the outputs of those agents are certainly worthy of humanistic interpretation. In other words, one can analyze the meaning of such texts, the motivations behind their creation, and the contexts in which they were created, insofar as they are parasitic on the motivations and meanings, as well as the contexts and cotexts, of such human agents—which includes the profit motives and ethical qualms of the corporate agents that spearheaded their creation.

Even setting aside intentionality, derivative or original, such texts were selected by minimizing cross-entropy loss or maximizing a reward signal, and by means of all the other modes of sieving and serendipity, and generativity more broadly, that went into their creation. (One could even argue that any process that minimizes or maximizes a function—and thereby takes it to an extreme—involves a glimmer of telos, if not a speck of intentionality. The second law of thermodynamics, as reformulated by Gibbs, is in a certain sense the origins of desire.) And so, as for any other living kind, one can offer a genealogy of their coming to be: the sorts of conditions and forces, from the size of silicon chips to the strivings of speculative capital, that contributed to their emergence.

If it is satisfying for someone to interpret such texts, however superficially or deeply, does it matter if they are the product of an aesthetic intention or existential motivation, however unconscious, any more than any other text one is compelled to interpret without reference to an authorial intention? Hermeneutics has long been done without reference to authors.

Indeed, the process often works the other way, and reciprocally so. Just as we can learn about a "person"—their identity, interests, and origins—by reading what they wrote, we can interpret what they wrote by reference to who they are as people. Bootstrapping processes may occur whereby people project not just personhood but also particular personalities onto large language models (given what they write, and how they respond). And then, by reference to such modes of personhood, people will reinterpret what such agents have written and how they respond. And those people may then prompt such personifications in new ways, looking for confirmation of their projections. Our understanding of Jesus and other (mainly) textually present people is not too different in its construction—if only as prompted through prayers. And so not just sects but also whole societies may performatively prompt such machinic persons into being—and thereby make not just their interpretations of texts but also their projections of personhood true, or at least true enough for the people—and "persons"—in question. And so yet another way for the world to be re-enchanted.

Indeed, setting aside any specific text it produces, the generative capacity of any language model is itself worthy of interpretations—no less than any dictionary or grammar, by linguistic analysis or otherwise. Think, for example, of the celebrated claims of the Italian humanist Giambattista Vico regarding the importance of Homeric texts: they contain "models or ideological portraits which form mental dictionaries of the ancients." Just add the words *grammar* and *pragmatics* to *dictionaries* in this quote, and change *ancients* to *moderns* (plus any prefix you might desire), and you are ready to plumb the generative depths of language models. In other words, language models are an incredible resource for studying the values of the people who wrote the texts they were trained on (not to mention the interests of those who fine-tuned the models to satisfy their alignment criteria). Such values include a collectivity's model of the world and all that it contains—however biased, irrational, or culture-bound. In a certain sense, however Borgesian, language models do not just contain all the texts that were written by a people, they contain all the texts that could have been written by that people given their worldview.

And, of course, this essay is essentially an interpretation and/or genealogy of language models per se, as well as a guide for how to approach the interpretation of any text they might generate, not to mention the motivation and interests of the agents (A_v, A_θ, A_p) that had a hand—or at least a say—in their creation.

10.
The Problem with Alignment

This chapter returns to some of the issues raised in the introduction, while reviewing and reworking key themes of this essay. I review various senses of alignment, as this term applies to semiotic processes, and compare them with the alignment of machinic parameters and human values. I then discuss and critique the alignment problem—the idea that it may be difficult, if not impossible, to make machinic behavior align with human values in truly important ways, given the open-endedness of the world and the uncertainty of the future.

Alignment

In everyday English the term *alignment* has many related senses. It may refer to the proper adjustment of components (in a system) for appropriate functioning (of that system). It can refer to an agreement, or alliance, between two or more parties. It can refer to the ground plan of a railroad or highway system (as opposed to the profile). And, of course, it can refer to entities being in a line, or, perhaps more frequently, to entities being in the appropriate relative positions, such that, wherever they happen to be, they face the same direction. In other words, it often refers to the orientation or stance of agents as opposed to their position. Are they directed to the same objects? Are they guided by the same ends?

With all this in mind, it is easy to envision various kinds of *semiotic alignment*. This may turn on representations, such as beliefs and assertions, aligning with the world (insofar as they are true). It may turn on the world coming to align with performative utterances, insofar as the latter are felicitous—and hence not just appropriate in context but also transformative of context. In other words, just as signs can come into alignment with objects, objects can come into alignment with signs. This may turn on interpretant-object relations corresponding to or coming into alignment with sign-object relations.

For example, does the interpreter come to look in the direction that the signer is pointing? Does the addressee come to believe—or not—what the speaker is saying? Is some kind of intersubjective agreement between agents achieved? And this may turn on semiotic agents having, or at least coming to have, the same values, interpretive grounds, or guiding principles. For example, do the agents share, or come to share, a set of conventions or a model of causal relations? Are their ontologies and partonomies in agreement? Do they have similar preference hierarchies or evaluative standards? Do they agree on what constitutes good alignment criteria? In short, semiotic processes and their conditions and consequences are easily framed in terms of different modes of alignment: between objects and signs; between sign-object and interpretant-object relations; and between the values of signifying and interpreting agents.

These ideas, in a slightly extended sense, have already been used to frame the relation between different kinds of semiotic agents. Recall figure 5, which showed how machinic interpretants (or "responses") can be brought into alignment with human signs (or "prompts"), insofar as machine parameters (θ) are brought into alignment with human values (V). It is also relatively easy, however anxiety-provoking, to add corporate agents, be they corporations or states, into the mix, for values and parameters can also be made to align with—as well as *counter-align* against—power and profit (*P*).

Problems with Alignment

The alignment of machine behavior with human interests is so important among AI ethicists that it has its own name: *the alignment problem*. More carefully, such a problem might be formulated as follows: Will the systems that we design and train ultimately do what we want and expect? And how do we ensure that they behave accordingly in an open-ended world and uncertain future? Phrased another way: How do we make sure that machine behavior (and thus the parameters that guide it, as well as the algorithms that underlie it) aligns with human values? And not just for now, but for all time, come what may?

Many futurists, ethicists, and experts in artificial intelligence have pondered these questions. And the extended discussion in chapter 4 regarding reinforcement learning with human feedback and the training of reward models showed one way that this challenge is being met, at least in a relatively circumscribed domain of satisfying users' intentions and corporate interests when responding to prompts.

Rather than delve further into the large, speculative, and unresolved literature around this topic, I pose five other problems—tangentially related to the alignment problem—that are, if not more pressing, at least more in line with the arguments of this essay.

First is what might best be called the *de-alignment problem*: with the advent of large language models and more and more sophisticated forms of artificial intelligence, the capacity of humans to create, share, and improve their values—which is perhaps the true generative endowment of the human species—may be weakened. And this diminishment is due, at least in part, to interference by noise and interception by enemies insofar as public discourse and private conversations come to be more and more mediated, and thus affected and directed, by what are essentially weaponized chatbots.

Second, and closely related to the first, is the *realignment problem*: human values may come to be more and more mediated by machine parameters (rather than vice versa), which may themselves be mediated by corporate agents with dubious and selfish, if not outright malicious, values and interests.

Third is the *provincial-value problem*: just as not all human voices are in the training corpus, not all human values determine machine parameters. So whose values were machines aligning with in the first place? And who gets to determine whose values machine parameters will align with in the future? In other words, never mind whether machinic parameters will align with human values; the question is which values, to what ultimate end, how we could know, and who should decide.

Fourth is the *posthuman problem*: machines might easily be trained to align with human values; the problem is that human values, whichever collectivity they happen to come from, may not be all that great to begin with—at least when the ultimate repercussions of value-guided semiotic processes (and hence human-specific modes of attention, inference, and action)—are examined on larger scales.

Fifth is the *Pandora problem*, which requires some explanation. As was shown in the discussion of dynamic generativity, how a tool can be used (dynamically) and how a tool may be used (deontically) are worlds apart. In other words, everything is reducible to its affordances: what is physically possible to do with it rather than what is normatively appropriate. This means that to rein in any machinic agent, a particularly powerful tool, we have to rein in all human and corporate agents from now on, insofar as they might be prone to abuse that agent's generative potential. In effect, every powerful new technology needs to be policed forever after to ensure appropriate usage—and such modes of surveillance and control ultimately may be worse for our collective existence than the modes of misuse they were designed to stop.

Finally, there is the *real alignment problem*: unchecked wealth inequality and resource extraction—and thus social hierarchies and environmental degradation—already pulled humans out of alignment with each other and with the earth (and most other living kinds). In other words, A_V is wildly out of alignment not just with itself but also with its true generative matrix, A_E, understood as the mother ship of all agency. So all this attention to large language models and generative AI is a massive distraction from the really pressing issues that currently hurt life on earth. Indeed, given the resources they consume, the conversations and debates they degrade, and the social relations they elide and strain, they are only adding fuel to the fire.

To return to the introduction, ChatGPT and the like are what we might call *hyperagents*: agents imbued with excessive hype relative to other agents. Such unjustified attention may be due to the fact that large language models seems to be coming for the jobs of the writing—or at least chattering—classes. Such people are precisely the ones who currently—but perhaps not for long—write articles, books, blog posts, screenplays, and tweets. So the simplest interpretant of this essay is that it is nothing but a symptom of the author's anxiety in the face of his own obsolescence.

But hopefully my arguments have offered more than that, thereby affording a wider range of interpretations. By focusing on the conditions for and consequences of techno-horizons, I have tried to cut through some of the bullshit that is espoused by large language models (and their makers, masters, and marketers), however eloquent and enchanting they may seem. And by bringing to light some of the more subterranean ways that values, parameters, and profits—as guiding principles—ground inference, action, intuition, and affect, I have sketched some of the ways that semiosis and sociality may be radically realigned in the facelessness of our new interlocutors.

References

This essay was first presented as a keynote address in May 2023, at the Technolinguistics in Practice: Socially Situating Language in AI Systems conference held in Siegen, Germany. Siri Lamoureaux, the key organizer, as well as Michael Castelle, Evan Donahue, Ilana Gershon, Yarden Skop, Alicia Fuentes-Calle, and Mark Dingemanse, offered very helpful feedback. For the stochastic parrot critique of language models, herein critiqued, see the important work of Emily Bender and colleagues. For more on discursive scaffolding, see *Cooperative Interaction* by Charles Goodwin. For classic work on entextualization, see the essays in *Natural Histories of Discourse*, edited by Michael Silverstein and Greg Urban, as well as the article by Richard Baumann and Charles Briggs, *Poetics and Performance as Critical Perspective on Language and Social Life*. Regarding the projection of agency in relation to discursive interaction, see *The Ontology of Action* by Nicholas J. Enfield and Jack Sidnell, and *Agency in Language* by Alessandro Duranti. Regarding reflexivity, semiotics, and subjectivity, see *Talking Heads: Language, Metalanguage, and the Semiotics of Subjectivity* by Benjamin Lee. For more on magic, see the work of Graham M. Jones. For a very different, but arguably allied take on gens, see jointly authored work by Laura Bear, Karen Ho,

Anna Lowenhaupt Tsing, and Sylvia Yanagisako. For a resonant work on generativity, see *Cultural Poesis: The Generativity of Emergent Things* by Katie Stewart. For more on alignment, see *The Alignment Problem* by Brian Christian. The originary work on transformers was *Attention is All You Need* (2017), jointly authored by researchers at Google. For a deep dive into large language models, and especially the GPT series, see the following articles produced by researchers at OpenAI: "Improving Language Understanding by Generative Pretraining" (2018), "Language Models are Unsupervised Multitask Learners" (2018), "Language Models are Few Shot Learners" (2020), "Deep Reinforcement Learning with Human Preferences" (2023), and "GPT-4 Technical Reports" (2023). For a step-by-step guide to building your own language models, the machine learning guru Andrej Karpathy has a wonderful YouTube series, *Building Makemore* (and much more besides). Many thanks to Matthew Engelke and Connor Martini for their illuminating and transformative feedback, as well as to Kamala Russell, Robert Meister, Terra Edwards, Jonathan Beller, Andrew Carruthers, and Julia Zrihen for inspiring suggestions.

Also available from Prickly Paradigm Press: